T0348750

Planar Cell Polarization During Development

ADVANCES IN DEVELOPMENTAL BIOLOGY

Volume 14

Series Editor

Paul M. Wassarman
Mount Sinai School of Medicine
Mount Sinai Medical Center
New York, New York

Editorial Board

PLANAR CELL POLARIZATION DURING DEVELOPMENT

Editor

Marek Mlodzik

Brookdale Department of Molecular, Cell, and Developmental Biology
Mount Sinai School of Medicine
New York, New York

2005

ELSEVIER

AMSTERDAM • BOSTON • HEIDELBERG • LONDON • NEW YORK • OXFORD
PARIS • SAN DIEGO • SAN FRANCISCO • SINGAPORE • SYDNEY • TOKYO

Elsevier
525 B Street, Suite 1900, San Diego, California 92101-4495, USA
84 Theobald's Road, London WC1X 8RR, UK

This book is printed on acid-free paper.

For all information on all Elsevier publications
visit our Web site at www.books.elsevier.com

ISBN-13: 978-0-44-451845-3
ISBN-10: 0-44-451845-2

Transferred to Digital Print 2008

Printed and bound by CPI Antony Rowe, Eastbourne

Contents

List of Contributors

Paul N. Adler   Biology Department, University of Virginia, Charlottesville, Virginia 22903

Kwang-Wook Choi   Department of Molecular and Cellular Biology, Program of Developmental Biology, Department of Ophthalmology, Baylor College of Medicine, Houston, Texas 77030

Alain Dabdoub   Section on Developmental Neuroscience, National Institute on Deafness and Other Communication Disorders, National Institutes of Health, Bethesda, Maryland 20892

Jason R. Jessen   Department of Biological Sciences, Vanderbilt University, Nashville, Tennessee 37235

Matthew W. Kelley   Section on Developmental Neuroscience, National Institute on Deafness and Other Communication Disorders, National Institutes of Health, Bethesda, Maryland 20892

Janghoo Lim   Department of Molecular and Cellular Biology, Program of Developmental Biology, Department of Ophthalmology, Baylor College of Medicine, Houston, Texas 77030

Marek Mlodzik   Mount Sinai School of Medicine, Department of Molecular, Cell and Developmental Biology, New York, New York 10029

Mireille Montcouquiol   Section on Developmental Neuroscience, National Institute on Deafness and Other Communication Disorders, National Institutes of Health, Bethesda, Maryland 20892

Petra Pandur   Abt. Biochemie, Universität Ulm, Albert-Einstein-Allee 11, 89081 Ulm, Germany

Amit Singh   Department of Molecular and Cellular Biology, Program of Developmental Biology, Department of Ophthalmology, Baylor College of Medicine, Houston, Texas 77030

Lilianna Solnica-Krezel   Department of Biological Sciences, Vanderbilt University, Nashville, Tennessee 37235

David Strutt   Centre for Developmental Genetics, Department of Biomedical Science, University of Sheffield, Western Bank, Sheffield, S10 2TN, United Kingdom

Helen Strutt   Centre for Developmental Genetics, Department of Biomedical Science, University of Sheffield, Western Bank, Sheffield, S10 2TN, United Kingdom

Preface

The establishment of planar cell polarity is one of the last frontiers of developmental biology. How individual cells hundreds of cells apart acquire the same polarization within an epithelial field, or how mesenchymal cells establish a uniform polarization prior to their intercalation, is a fascinating biological problem. Much progress has been made in recent years, although the molecular aspects of the process are still far from being understood. Strikingly, the mechanisms appear largely conserved across all metazoa. In this book, we are trying to provide an overview of the current knowledge and features of planar cell polarization (PCP) in different contexts. As this book is being completed several papers have appeared that further extend our understanding of PCP or challenge the views presented here. Thus by no means do we claim this compilation is either complete or fully up-to-date. Nevertheless, we hope it provides a general overview and attempts to cover many different tissues and models of PCP establishment. We apologize to the research areas and view points that have not been included here for whatever reason.

Different aspects of cellular polarization

The establishment and maintenance of cellular polarization is of general importance for the proper development and functioning of most organs and all organisms. Cells in many distinct contexts show different aspects of polarization. Apical-basolateral polarity, for example, is evident in all epithelial cells. Additionally, some epithelial cell types are polarized within the plane of the epithelium, perpendicular to the apical-basal axis. This type of polarity is generally referred to as tissue polarity or planar cell polarity (PCP). PCP has been most studied and is best understood in *Drosophila*, where all adult epithelial cuticular structures show tissue polarity features and are polarized within the plane. Genetic studies in *Drosophila* led to the identification of the first PCP genes some twenty years ago, this first set included also the *frizzled* (*fz*) gene. The study of PCP establishment has recently also been extended to vertebrates, with emphasis on the polarization of the sensory cells in the inner ear epithelium and the morphology and behavior of mesenchymal cells undergoing the morphogenetic process of convergent extension during gastrulation. The combination of the study of PCP establishment in *Drosophila* and vertebrates (in several organisms and tissues) has thus accelerated our insights into and understanding of the process. Nevertheless, much work remains to be accomplished before we can claim of at least partly understanding what is going on.

Whereas apical-basolateral polarity is often established simply through local extracellular clues like cell adhesion properties, PCP establishment requires long range, complex signal propagation in order to ensure that all cells in a tissue are precisely oriented. It is thought that the interpretation of the polarizing signal is mediated by a core Fz/PCP signaling cassette in all PCP responsive cells. The specific cellular interpretations of this signaling cassette, however, are very diverse, resulting in cytoskeletal rearrangements, changes in cell adhesion properties, reorientation of the mitotic spindle, or other modifications depending on the specific cell type involved. Additionally, PCP signaling effects these changes not only on individual cells, but also on large multicellular units, altering their polarity as a group with respect to the surrounding environment.

Cellular polarization is critical for almost all cell types and is often associated with diseases when disturbed. Epithelial cells and tissues require the apical-basolateral polarity to perform vectorial functions, including the directed transport of fluid and the directed secretion of specialized components. The reported functions of PCP in vertebrates include aspects of body hair orientation and skin development, polarization of the sensory epithelium in the inner ear, and the directed movement of mesenchymal cell populations during gastrulation. Other vertebrate PCP functions can easily be envisioned and/or already have been proposed, including the polarization of cilia in the oviduct and respiratory tract. I will try here to give first a brief account of the historical development of the field, and the follow this with general comments about the state of the PCP field and its many potential impacts on other signaling pathways and/or medical disease states.

The history of the tissue polarity field

The study of PCP originates from work in the fruit fly *Drosophila,* where it was initially called 'tissue polarity'. Elegant work pioneered by the group of Paul Adler has put the problem cleanly on the map some 20 years ago. This was accompanied by the genetic identification of other genes and the emergence of the first models, mostly through work by Paul Adler, David Gubb, and Peter Lawrence and their colleagues. The genetic characterization of *Drosophila* tissue polarity genes was soon followed by the first molecular cloning of a PCP gene in *Drosophila: frizzled.* Again this ground breaking work was performed by Paul Adler and colleagues (see also chapter by Paul Adler in this book). These initial insights were followed up by systematic genetic screens by several research teams in fruit flies and molecular cloning and analysis of several PCP factors. Major advances have thus been made in our understanding of the signaling circuitry and mechanistic aspects of PCP establishment in flies.

The analysis of PCP is now an important feature of developmental studies in many organisms. Starting in Xenopus and zebrafish, with the analysis of the process of convergent extension during gastrulation and neurulation, related processes where discovered and analyzed in vertebrates. Strikingly, the large genome wide forward genetic screens in the zebrafish identified several mutants affecting convergent extension, and following their cloning these turned out to be orthologues of several *Drosophila* PCP genes. Among these are both components of the core primary PCP group, as well as components of a Frizzled PCP specific signaling pathway.

The parallels and conservation of the PCP gene cassette are now extended also to vertebrate epithelia. In particular, the mammalian inner ear is a beautiful example of a neural epithelium with PCP features. Strikingly, the homologues of the fly PCP genes are showing up again when mutants with ear PCP defects are being analyzed (see chapter by Dabdoub et al. in this book). Importantly, some of the vertebrate homologues affect both the inner ear as well as convergent extension during gastrulation and neurulation, supporting the common mechanism of planar polarization not only across species but also between different polarized cell types and organs.

Recent work by Jeremy Nathans and colleagues (Guo, N., Hawkins, C., Nathans, J. 2004. *Frizzled6* controls hair patterning in mice. Proc. Natl. Acad. Sci. USA 101, 9277–9281) closes the circle of PCP features and beautifully demonstrates that the PCP principle can be extended to the mammalian epidermis. The whole field started with the discovery and analysis of *Drosophila frizzled* based on its phenotype in the fly cuticle (generally speaking the cuticle could be considered an equivalent to the mammalian epidermis as it represents the body 'skin') and the work by Guo et al. (2004) now shows that the mammalian epidermis requires also a Frizzled gene for its regular pattern. Even more striking is that the pattern defects in the mouse *Frizzled6* mutant are very much reminiscent of the phenotypic abnormalities of fly *frizzled* mutants. Thus, this is exciting not only for the genetic demonstration of the epidermal PCP features in mammals, but also (1) for the discovery that the same evolutionarily conserved protein family regulates this process from flies to mammals, and (2) that the defects observed in the respective mutants are virtually identical between flies and mammals, suggesting very similar principles at work in all contexts.

Same signaling cassette, but different read-outs

The conserved core PCP gene cassette is conserved. Nevertheless, as it is used in many different contexts in flies or vertebrates, the cellular read-outs are very distinct from tissue to tissue. The effector pathways include cytoskeletal organization, nuclear signaling or orientation of the mitotic spindle. These basic read-outs are again likely conserved from across the animal kingdom.

Cytoskeletal organization is likely the main 'target' of PCP in cuticular cells in Drosophila, the inner ear epithelium in mammals, or the skin in vertebrates (see also chapters by Adler and Dabdoub et al. in this book). A nuclear signaling response is prominent in all multicellular units, e.g., in the *Drosophila* eye (each ommatidium forming a unit, see chapter by Mlodzik in this book) or likely also in feather buds in birds. The PCP regulation of the orientation of mitotic spindles is known in *Drosophila* and *C. elegans*, and likely occurs in other animals as well. In addition, there are PCP controlled processes where the read-out is not yet obvious. These include the cellular behavior and movements during convergent extension in gastrulation and neurulation (see also chapter by Jessen et al. in this book). Although the PCP cassette is clearly required in this context, it is not yet determined what it really regulates in the mesenchymal cells as a cellular read-out. It is likely that other shared PCP read-outs will emerge in the near future.

Establishing signaling 'centers' and long-range PCP patterning

Despite the recent insight into the cellular short range interactions among the core PCP genes, the aspect of long range regulation of Fz-PCP signaling and the establishment of polarizing sources (that could regulate whole fields of cells) remain largely obscure. In particular, the fact that a PCP dedicated Wnt is still elusive in *Drosophila* raises many questions. Although several Wnt family members have been linked to PCP signaling in vertebrates (e.g., see chapter by Jessen and Solnica-Krezel and Dabdoub et al.), no evidence for a PCP specific *Drosophila* Wnt member exists todate. Thus, the long-range regulation of PCP signaling remains largely an unresolved problem.

Nevertheless, some progress has been made in *Drosophila* in this regard. In the fly eye the establishment of a presumed signaling source (the equator) has been quite extensively studied, and suggests that several signaling pathways lead to a restriction of expression of several factors, which could act as potential long range regulators of PCP or at least contribute to the long range regulation of Frizzled-PCP activity (see chapter by Singh et al.). The observation that several factors, implicated in PCP establishment upstream of Frizzled, are expressed in a graded manner within the developing eye disc might provide an insight and explanation into the long range regulation of PCP signaling. Most notably the proto-cadherins Fat and Dachsous, and the transmembrane protein Four-jointed have received attention in this context. Functional analyses have provided support for their general requirement and conserved function as long range regulators of PCP (see chapter by Strutt and Strutt). Interestingly, their graded expression is not restricted to the developing eye but is also observed in other tissues, e.g., during the polarization of the fly wing. Thus, although the jury is still out on this, Fat, Dachsous, and Four-jointed are attractive candidates for being long range PCP regulators, or at least to contribute to the long range regulation in this context.

Multiple Wnt/Frizzled signaling pathways and multiple ways to generate PCP

One striking shared feature of the analysis of Frizzled-PCP signaling is that it utilizes a distinct pathway from canonical Wnt/Frizzled signaling. The Frizzled receptor family has a major function in the reception and transduction of Wnt signals leading to the activation of the β-Catenin signaling pathway, commonly now referred to as canonical Wnt signaling. In contrast, the Wnt/Frizzled-PCP pathway does not affect β-Catenin signaling, and does not share any other components of the canonical pathway downstream of Dishevelled, which appears to be the only cytoplasmic component shared between the PCP and β-Catenin signaling pathways. The analysis and molecular dissection of the signaling specificity is an active area of research and likely to be critical in our understanding of either signaling outcome.

The Wnt/Frizzled signaling complexities do not stop with the PCP and β-Catenin pathways. Recent work has added several other potential effector pathways downstream of Wnt/Frizzled (see chapter by Pandur in this book). The ongoing analysis suggests an increasingly complex interaction network and potential cross-talk between these distinct effector pathways. The Wnt/Frizzled 'circuitry' is however getting even more complex with the discoveries of new non-Wnt ligands, e.g., Norrin (Xu, Q., Wang, Y., Dabdoub, A., Smallwood, P.M., Williams, J., Woods, C., Kelley, M.W., Jiang, L., Tasman, W., Zhang, K., Nathans, J. 2004. Vascular development in the retina and inner ear: control by Norrin and *Frizzled-4*, a high-affinity ligand-receptor pair. Cell 116, 883–895) and new non-Frizzled Wnt receptors, e.g., Derailed/ Ryk (Inoue, T., Oz, H.S., Wiland, D., Gharib, S., Deshpande, R., Hill, R.J., Katz, W.S., Sternberg, P.W., 2004. *C. elegans* LIN-18 is a Ryk ortholog and functions in parallel to LIN-17/*Frizzled* in Wnt signaling. Cell 118, 795–806; and Lu, W., Yamamoto, V., Ortega, B., Baltimore, D. 2004. Mammalian Ryk is a Wnt coreceptor required for stimulation of neurite outgrowth. Cell 119, 97–108). Thus although our understanding of Fz-mediated PCP establishment has increased dramatically warranting several reviews and this book, it appears that we have some far only been scratching the surface of the many complexities to be discovered in this context.

Lastly, although this book focuses on aspects of PCP establishment that are mediated by Frizzled signaling and the associated gene cassette, there is clearly more to planar cell polarization processes and establishment than just this pathway. Recent work in *Drosophila* embryogenesis has nicely shown that a polarization of cells within the embryonic ectodermal cell plane uses molecular mechanisms that do not rely on any of the components of the Frizzled-PCP gene cassette (Zallen, J.A., Wieschaus, E. 2004. Patterned gene expression directs bipolar planar polarity in *Drosophila*. Dev. Cell 6, 343–355; and Bertet, C., Sulak, L., Lecuit, T. 2004. Myosin-dependent junction remodelling controls planar cell intercalation and axis elongation. Nature 429, 667–671). Despite this surprising observation, the cellular aspects of the early cell intercalation during Drosophila germ band extension look very reminiscent of the convergent extension process during vertebrate development, which requires the Fz-PCP signaling pathway.

These observations again show that we know very little about PCP in general and there much more to be discovered in cellular polarization mechanisms. In addition, to trying to understand Frizzled mediated PCP with all its complications, obviously we have to look now also outside the "Frizzled signaling box".

MAREK MLODZIK

Planar polarity in the *Drosophila* wing

Paul N. Adler

Biology Department, University of Virginia, Charlottesville, Virginia 22903

Contents

1. Introduction

The cuticle of insects is dramatically decorated with a variety of polarized structures such as trichomes, bristle sense organs, and denticles. In any body region these
cuticular structures typically point in the same direction giving each body region a
tissue planar polarity (Lawrence, 1966). Research on *Drosophila* has pointed the way
towards understanding the cellular, molecular, and genetic basis for this (Gubb and
Garcia-Bellido, 1982; Adler, 2002). The function of the *frizzled* (*fz*) pathway results
in the consistent polarization of all cells in a body region. This pathway is sometimes
called non-canonical Wnt pathway (Veeman et al., 2003a) but in *Drosophila* planar
polarity there is no evidence for Wnt involvement (Lawrence et al., 2002). Hence in
this review I will use the term *fz* pathway. Recent data has shown that the homologous genes function to organize planar polarity in vertebrate gastrula undergoing

Advances in Developmental Biology
Volume 14 ISSN 1574-3349
DOI: 10.1016/S1574-3349(04)14001-5

convergent extension and in orienting the stereocillia of the vertebrate inner ear (Heisenberg et al., 2000; Darken et al., 2002; Goto and Keller, 2002; Mlodzik, 2002; Carreira-Barbosa et al., 2003; Curtin et al., 2003; Montcouquiol et al., 2003; Veeman et al., 2003a,b). Thus, this pathway appears to be conserved for the development of cell polarity within the plane of a tissue.

The wing is the largest appendage in *Drosophila* and at least the wing blade (not considering the hinge) is a quite simple structure. The adult cuticular wing is essentially a cast of the pupal wing, which consists of a one-cell-thick folded epithelial sheet. Most wing cells differentiate to produce a single distally pointing cuticular hair (Fig. 1). This gives the appendage as a whole a planar tissue polarity. This is also true for other appendages and most of the trunk. The distal direction of the hair is a consequence of the hair being formed at the distal-most part of the cell and growing out away from the cell (Wong and Adler, 1993). Thus, each individual cell on the

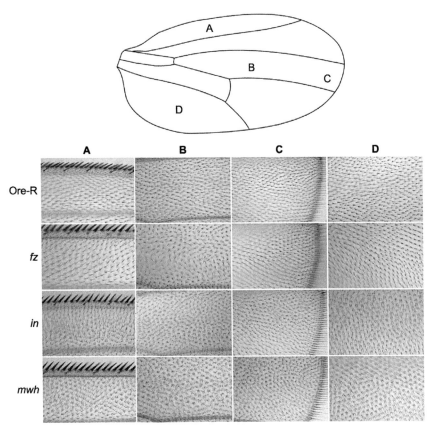

Fig. 1. The wing phenotypes of planar polarity mutants. Shown above is a drawing of a wing with four regions noted (ABCD). Below are micrographs from these regions from wild type (Oregon R) and *fz, in,* and *mwh* mutant wings. The micrographs for regions A, C, and D are from the dorsal surface of the wing and B is from the ventral surface of the wing. Hairs are smaller on the ventral surface and the mutant phenotypes are typically stronger.

wing is also polarized in the plane of the epithelium. In this review, I will refer to this as planar cell polarity.

2. Genetics of planar polarity genes

Forward genetics has identified a number of *Drosophila* genes important for the development of wing planar polarity (Figs. 1, 2, 3). One set of genes appears to function cooperatively in the polarization of wing cells. A key gene is *frizzled (fz)*, which encodes a serpentine plasma membrane receptor (Vinson et al., 1989) (this gene was the founding member of the *fz* family of Wnt receptors although there is no evidence for a Wnt ligand in wing planar polarity). Other central genes include *starry night (stan)* (also known as *flamingo (fmi)*) which encodes a protocadherin (Chae et al., 1999; Usui et al., 1999), *disheveled (dsh)* which encodes a PDZ domain-containing signal transducer (Klingensmith et al., 1994; Theisen et al., 1994), *prickle (pk)/spiny leg (sple)* which encodes LIM domain proteins that are alternative products of a complex gene (Gubb et al., 1999), *Vang Gogh (Vang)* (also known as *strabismus (stbm)*), which encodes a putative transmembrane protein (Taylor et al., 1998; Wolff and Rubin, 1998) and *diego (dgo)* which encodes an ankyrin-containing

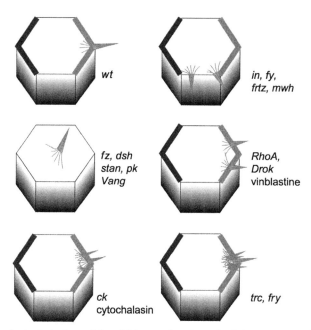

Fig. 2. Drawings of a typical wing cell in wild type and mutants (and drug treatments). The developing hair, the accumulation of *fz/dsh* at the distal side of the cell, and the accumulation of *pk/Vang(stbm)* are shown. The lack of an effect on *fz/dsh/pk/Vang(stbm)* accumulation in *Drok, ck* and *trc* group mutants has not been established and is presumed to be the case due to the hair forming at the distal side of the cell.

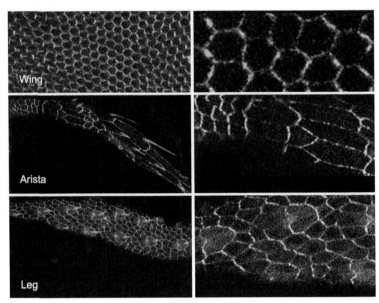

Fig. 3. Shown are both low and high magnification micrographs of pupal wing, leg, and arista cells that both express *fz*-GFP and are stained for F-actin using Alexa568-phalloidin (red). Only GFP is shown in the high magnification images. In these micrographs, *fz*-GFP is visualized directly without antibody staining. Note the distinctive zigzag *fz* pattern in the wing and that a similar pattern is seen in arista and leg, albeit modified to fit the cellular geometry of these cell types. In all of these cell types the *fz*-GFP is found in an uneven and clumpy pattern in the regions where it accumulates. Note how in the wing cells the actin-rich hair is formed at the edge of the cell, perhaps in direct contact with the accumulated *fz*. In the arista however, the laterals form some distance from the accumulated *fz*-GFP. Note that *fz*-GFP accumulated asymmetrically in the developing bristle socket cells on the leg. The arista and leg images are modified versions of images from He and Adler, 2002. (See Color Insert.)

protein (Feiguin et al., 2001). I will refer to these as the *fz* group. A number of proteins have been implicated in regulating the activity of the *frizzled* group. These include the protocadherins *fat* (*ft*) (Mahoney et al., 1991) and *dachsous* (*ds*) (Clark et al., 1995), the secreted protein *four jointed* (*fj*) (Zeidler et al., 2000), and the *atrophin* transcriptional regulator (Fanto et al., 2003). At least two distinct groups of downstream genes have been identified. One group includes genes such as *inturned* (*in*), *fuzzy* (*fy*), *fritz* (*frtz*), and *multiple wing hairs* (*mwh*), where mutations result in both altered polarity and the formation of multiple hairs (Gubb and Garcia-Bellido, 1982; Wong and Adler, 1993; Adler, 2002). The second group includes genes such as *RhoA* (Strutt et al., 1997), *Drok* (*Drosophila* Rho kinase) (Winter et al., 2001), *crinkled* (*ck*) (which encodes a myosin VII), and *zipper* (*zip*) (which encodes a myosin II) which are good candidates to be direct regulators of the cytoskeleton (Turner and Adler, 1998; Kiehart et al., 1999; Strutt, 2001a; Winter et al., 2001; Adler, 2002). Mutations in these genes result in the formation of multiple hair cells but do not substantially alter polarity. Hence, these genes appear to function in planar cell polarity but not planar tissue polarity.

The phenotypes of null mutations in these genes are informative and often surprising to those who do not work in the field. An interesting point is that many of the genes are not essential. Strong mutant alleles, however, result in weak and inactive flies that would likely not survive for long outside of the lab. In a *fz* mutant the vast majority of wing cells still form a single hair but most do not point distally (Fig. 1). Over much of the wing the general polarity pattern is reproducible from one fly to another (Gubb and Garcia-Bellido, 1982). Thus, in the D and E cell region of a *fz* wing (located in the posterior) hairs point posteriorly toward the margin and not distally as in wild type (Gubb and Garcia-Bellido, 1982; Wong and Adler, 1993; Adler et al., 2000). Neighbors remain well aligned with one another. Interestingly, in the most distal part of the wing, polarity is similar to wild type (Fig. 1). Thus, some planar polarity remains and is independent of *fz* pathway function. Mutations in *stan/fmi*, *Vang/stbm*, and *dsh* result in very similar wing phenotypes both with respect to the polarity phenotype and the number of multiple hair cells (Adler et al., 2000). Several of these genes (*fz, pk, Vang/stbm*) are not essential, while others are (*stan/fmi, dsh*). This presumably is due to the essential genes functioning non-redundantly in additional pathways. Mutations in *in, fy, mwh*, and *frtz* also result in a very similar abnormal polarity pattern, albeit perhaps a bit more severe (Fig. 1; Wong and Adler, 1993; Adler et al., 2000). None of these four genes are essential. Mutations in these genes differ from the *fz* phenotype primarily due to many cells forming more than one hair (Gubb and Garcia-Bellido, 1982; Wong and Adler, 1993; Adler, 2002). The multiple hair cell phenotype of *mwh* is substantially more severe with most cells forming three hairs or more (Fig. 1) (Gubb and Garcia-Bellido, 1982; Wong and Adler, 1993). For all of the genes noted above the strength of the multiple hair cell phenotype varies as a function of wing location (Fig. 1). The significance of this is unclear. All of the genes noted earlier as potential direct regulators or components of the cytoskeleton are essential presumably due to other cellular functions for these proteins.

3. Asymmetric accumulation of planar polarity proteins

Important recent progress has been derived from findings that the protein products of tissue polarity genes accumulate at either the proximal (*pk/sple* (Tree et al., 2002), *Vang/stbm* (Bastock et al., 2003)), the distal (*fz* (Strutt, 2001b) and *dsh* (Axelrod, 2001; Shimada et al., 2001)), or both the proximal and distal (*stan/fmi* (Usui et al., 1999) (*diego* has also been found to localize asymmetrically but the side is uncertain (Feiguin et al., 2001))) sides of wing cells (Figs. 2, 3). This process requires the function of all of the proteins thus far shown to accumulate asymmetrically. Thus, in a *fz* mutant none of the proteins accumulate asymmetrically. However, it appears that the function of each of these proteins is not equivalent. For example, mutations in *stan/fmi* result in a failure of apical accumulation of all of the other proteins while mutations in *dsh* results in *fz* and *stan/fmi* being evenly distributed around the apical cell periphery. It has been suggested that *stan* is responsible for recruiting *fz* and *Vang/stbm* to the apicolateral adherens junction (Bastock et al., 2003). The function of these membrane proteins is then required to recruit *dsh*, *pk*, and *dgo* to the membrane. This topic has recently been reviewed in depth (Strutt, 2002). The cell biological basis (e.g., directed transport on microtubules) for the asymmetry is unclear. It is worth noting that a similar asymmetric localization of *fz* pathway proteins is also seen in epidermal cells in other body regions such as the antenna and leg (Fig. 3) (He and Adler, 2002). Interestingly, *stan/fmi* was reported not to accumulate asymmetrically in bristle sense organ precursor cells (Lu et al., 1999), although asymmetric accumulation of *fz* was seen in developing bristles (He and Adler, 2002). This point needs to be re-examined in more depth. The situation in the eye is both similar and different from that in the wing. In the eye the *fz* pathway proteins accumulate asymmetrically but this is not uniform across the tissues. This is seen in only a subset of cells (most notably the R3 and R4 photoreceptors) and, consistent with previous results, the R3 and R4 cells are key for specifying eye planar polarity (Zheng et al., 1995; Das et al., 2002; Strutt et al., 2002; Yang et al., 2002).

4. Linking morphology to planar polarity

The consistent proximal-to-distal planar polarity on appendages requires a mechanism that links local planar polarity to the overall body plan. This is likely to work by spatially regulating the activity of the *fz* pathway. One class of models to explain this relies on the localized production of global polarity signals. There are two obvious candidate locations for producing such signals on the wing. One is the wing hinge and the other is the wing margin (or distal tip of the wing). Many mutations are known that result in the loss of small or large segments of the wing margin and none of these have a wing planar polarity phenotype. In extreme examples (such as *vestigial* mutations) much of the wing blade is lost. The lack of a polarity phenotype in such wings indicates that the wing margin cannot be the source of a global polarity signal unless it is spatially redundant or the signaling takes place early in disc development. Surgical manipulations have removed distal wing tissue from contact with proximal wing regions during the 14 hours prior to wing hair morphogenesis (this encompasses the temperature-sensitive period for *fz* function in wing planar polarity (Adler et al., 1994; Strutt and Strutt, 2002)). Such treatments did not alter the polarity of the distal wing cells (Turner and Adler, 1995; Adler et al., 2000). Thus, if there is any global signaling it likely takes place earlier in development.

What properties are to be expected in genes involved in coupling planar polarity to overall morphology? One prediction is that mutations that inactivate components of such a system will not inactivate the *frizzled* pathway, although they might be expected to disrupt normal tissue planar polarity. Several different approaches have been used to assay for *fz* pathway function under conditions of abnormal polarity as a way to identify planar polarity regulators. These include the overall mutant polarity pattern (as noted above, mutations that inactivate the *fz* pathway result in a stereotypic abnormal polarity pattern) (Adler et al., 2000), the directional domineering nonautonomy of *fz* and *Vang/stbm* clones (Adler et al., 1998, 2000) and the polarized accumulation of planar polarity proteins (Strutt and Strutt, 2002; Tree et al., 2002;

Ma et al., 2003). There is substantial support for the hypothesis that the activities of the *ds, ft,* and *fj* genes regulate *fz* pathway signaling, although there are differences in the detailed interpretations. The most complete data is for *ds,* which fits all of the criteria mentioned above. It has been proposed that *ft, ds,* and *fj* function to regulate the *fz* pathway and are responsible for linking planar polarity to overall body morphology (Strutt and Strutt, 2002; Yang et al., 2002; Ma et al., 2003). One model utilizes a distal-to-proximal gradient of Fj and a proximal-to-distal gradient of *ds* acting through *ft* to provide the overall orientation (Ma et al., 2003). The extent to which there is a proximal-to-distal gradient of *ds* expression is debatable (Strutt and Strutt, 2002) however, a "marginal gradient" could be sufficient if these genes functioned only to provide a rough proximal-to-distal gradient of *fz* activity as a first step in polarity establishment as has been proposed. There is also evidence for *prickle* mutations altering the direction of *fz* pathway signaling (Adler et al., 2000). The alleles where this was seen were mutations that altered the balance of two of the products of this complex locus *pk* and *sple* (Gubb et al., 1999). Mutations that inactivate both appear to block *fz* pathway function (Gubb et al., 1999).

5. What is the basis for the domineering nonautonomy of *fz* mutations?

The ability of a clone of *fz* mutant cells to alter the hair polarity of neighboring wild type cells (Fig. 4A) was one of the early landmark properties of wing planar polarity (Vinson and Adler, 1987). This domineering nonautonomy was originally described as being directional as cells which were distal but not proximal to the clone were affected. A later interpretation was that the presence of *fz* mutant cells causes surrounding cells to repolarize so that hairs tend to point toward the clone (Adler et al., 1997). Cells proximal to the clone do not appear to be affected as they always pointed in the direction of the clone (Fig. 4A). This interpretation is consistent with the observation that hairs point from cells of higher *fz* levels toward cells of lower levels. Clones mutant for *Vang/stbm* have a complementary phenotype and tend to point away from the clone (Taylor et al., 1998). Clones of other *fz* pathway mutants such as *pk* or *stan/fmi* show very weak domineering nonautonomy—at least an order of magnitude weaker than that seen for *fz* or *Vang/stbm* (Gubb et al., 1999; Adler et al., 2000). Two distinct classes of models have been proposed to account for these results and to explain planar polarity on a local level. One group of models suggests that there is a gradient of *fz* activity across the wing and this results in the gradient production of a locally diffusible secondary signal (Adler et al., 2000). This secondary signal polarizes the wing cells resulting in hairs pointing from higher to lower levels of the signal (Fig. 4C). A clone of cells lacking *fz* function results in a local depression in the concentration of the secondary signal, which in turn results in cells forming hairs that point toward the clone. Clones of cells lacking *Vang* activity would produce an excess secondary signal leading to cells surrounding the clone producing hairs that point away from the clone. Such models nicely explain the complementary nonautonomy of *fz* and *Vang/stbm* clones and the observation that hairs point down a *fz* activity gradient, but they do not provide a molecular mechanism. Nor do they

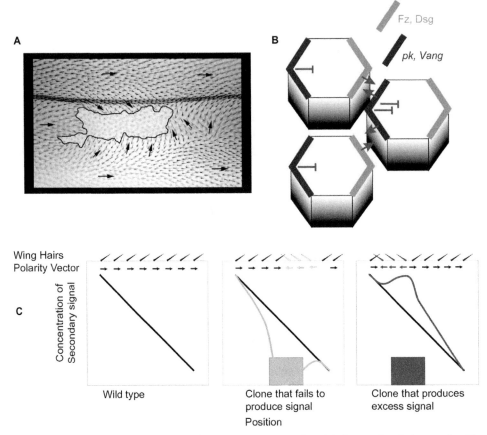

Fig. 4. Cell autonomy/non-autonomy of planar polarity mutants. Panel A shows a *fz strb* clone outlined in black (*strb* cells form small hairs that do not photograph well). Note that the hairs produced by neighboring wild type cells appear to be "attracted" to the clone. In panel B the cell-by-cell feedback model is drawn. The accumulation of *fz/dsh* on the distal side of the cell promotes the accumulation of *pk/Vang* (*stbm*) on the proximal side of the adjacent cell (and vice versa). The accumulation of *pk/Vang(stbm)* locally inhibits the recruitment of *dsh* to the membrane (Tree et al., 2002). Panel C shows how the secondary signal model can explain the reciprocal domineering nonautonomy of *fz* and *Vang/stbm* clones (Taylor et al., 1998). The presence of a *fz* clone results in a failure to produce the secondary signal and a local deficit that spreads beyond the clone. The presence of a *Vang/stbm* clone results in excess production of the secondary signal and this results in a local surplus that also spreads beyond the clone.

explain the weak domineering nonautonomy seen with mutations in *pk* and *stan/fmi*. Similar models have also been suggested for explaining *fz* pathway function in the eye (Zheng et al., 1995; Tomlinson and Struhl, 1999; Zeidler et al., 2000). An alternative model proposes that the asymmetric accumulation of planar polarity proteins is propagated on a cell-by-cell basis due to the interaction between distally and proximally localized proteins (Tree et al., 2002). This model is attractive in that it incorporates a plausible feedback system (always a plus in biology) and there is experimental data that argues that the presence of *pk* on the proximal side of a cell promotes the accumulation of *fz* and *dsh* on the distal side of the neighboring cell (and vice versa) (Tree et al., 2002). The principal drawback of this model is that it does not provide an obvious explanation for the order of magnitude difference in the degree of domineering nonautonomy between *fz* and *Vang/stbm* on one hand and *stan/fmi, dsh*, and *pk* on the other as all of the proteins encoded by these genes are all required for the asymmetric accumulation of the others. It does however, provide an attractive rationale for the weak domineering nonautonomy of genes such as *pk* and it provides an explanation for the corequirement of all of these proteins for the localized accumulation of the group. It can also explain the co-uneven localization along the proximal/distal cell boundary that is often seen. It is also effective in explaining why the directed expression of most, if not all, planar polarity proteins results in a repolarization of hairs. For some genes, such as *fz,* polarity goes from high toward lower levels (Adler et al., 1997), while for others, such as *stan/fmi,* hair polarity points from low toward high (Usui et al., 1999). Such observations fit nicely into the cell-by-cell feedback model but not into the secondary signaling model, at least for genes such as *stan/fmi* that are largely cell autonomously acting. It is possible, but not parsimonious, that both models could be correct. There is evidence for *fz* having both cell autonomous and non-autonomous functions (Vinson and Adler, 1987) and evidence has indicated that the autonomous function is later than the nonautonomous function (Strutt and Strutt, 2002). Thus it is possible that the hypothesized secondary signaling system functions prior to the feedback system that polarizes individual cells. The weak domineering nonautonomy of mutations in genes such as *stan/fmi* would be due to the feedback system. The building complexity of what is known about the system points out the need for quantitative and rigorous modeling to evaluate the different possibilities and for in vivo observations of developing planar polarity.

6. How does the asymmetric accumulation of *fz/dsh* and *pk/Vang(stbm)* regulate the cytoskeleton?

The accumulation of *fz* and *dsh* at the distal edge of the cell and *pk* and *Vang* at the proximal side of wing cells presumably produces a polarized signal that results in the cytoskeleton being activated to produce a hair at the distal-most part of the cell. There are several possibilities to consider. First the cytoskeleton could be activated distally or inhibited proximally (or both). Distal activation could work by direct interaction of cytoskeleton regulators with the *fz* and *dsh* proteins localized at the distal edge (Fig. 5). This is plausible for the wing where the formation of the hair is closely juxtaposed to the accumulation of *fz/dsh*. Observations on developing aristae make this unlikely to be a general mechanism (He and Adler, 2002). The *fz* pathway regulates the position for lateral development along the proximal/distal axis of aristae cells. These are large elongated cells and there is no overlap between the location for lateral initiation and the accumulation of *fz/dsh* (Fig. 3). At least in this cell type the

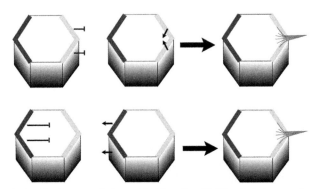

Fig. 5. How does the localized accumulation of *fz/dsh* and/or *Pk/Vang(stbm)* cause the cytoskeleton to be activated at the distal side of the cell to produce a hair there? Four distinct possibilities are shown. *fz/dsh* could activate the cytoskeleton intracellularly or inhibit hair formation in the adjacent cell. Similarly, *pk/Vang(stbm)* could inhibit hair formation intracellularly or promote it in the adjacent cell.

asymmetric accumulation of *fz/dsh* could activate the cytoskeleton indirectly. An alternative mechanism for the distal activation of the cytoskeleton in wing cells would be to have the proximally localized *pk* and *Vang(stbm)* proteins signal and activate the cytoskeleton on the distal side of the neighboring proximal cell (Fig. 5). This would be an intercellular activation of the cytoskeleton and predicts that mutations that disrupt this would show domineering nonautonomy with a high cellular penetration at the proximal border of clones and rescue of clone cells at the distal clone boundary. This has not been reported but it might have been missed. An alternative hypothesis is that the proximal accumulation of *pk* and *Vang/stbm* results in an inhibitory signal and in that way indirectly promotes hair initiation at the distal edge or that the distal accumulation of *fz/dsh* inhibits hair formation at the proximal side of the neighboring cell (Fig. 5). Determining if it is the distal or proximal protein complex that is responsible for restricting hair formation to the distal side of the cell is made difficult due to our current inability to inactivate the distal and proximal protein complexes independently.

7. Downstream of *fz*

What gene products transduce the signal of localized *fz/dsh* and/or localized *pk/Vang(stbm)* to the cytoskeleton? The products of the inturned-like genes are strong candidates to function as planar polarity effectors as their function is essential for the localized accumulation of *fz/dsh* and *pk/Vang(stbm)* to affect the location used for hair initiation. Mutations in *in* and *fy* are epistatic to both gain of function and loss of function mutations in *fz* group genes (Wong and Adler, 1993; Lee and Adler, 2002). At the current time it is not known if *in* or *fy* become asymmetrically localized within wing cells. This would be predicted by a number of models but it is worth noting that the overexpression of *in* or *fy* does not produce a gain of function wing hair phenotype (Park et al., 1996; Collier and Gubb, 1997). Based on the examples from the *fz*-like genes, where overexpression leads to a loss in the normal restricted localization, it is reasonable to hypothesize that *in* and *fy* will not need to be localized to a small part of a wing cell to function (Krasnow and Adler, 1994; Usui et al., 1999; Axelrod, 2001; Strutt, 2001b; Bastock et al., 2003). Definitive evidence will require direct localization of these proteins.

In developing pupal wing cells, *fz* and *dsh* accumulate across about one-third of the cell periphery while hair formation is restricted to a substantially smaller region (perhaps one-fifth of this region) (Fig. 3). Hence, there is a need for a further restriction in the location for hair formation. This seems likely to involve the function of the cytoskeleton itself as mutations in cytoskeletal proteins (e.g., myosin II) and known cytoskeletal regulators (e.g., *Drok*) result in the formation of multiple hairs along the distal side of the cell (Strutt et al., 1997; Turner and Adler, 1998; Kiehart et al., 1999; Winter et al., 2001). Consistent with this hypothesis these phenotypes are phenocopied by treatment with drugs that antagonize either the actin or microtubule cytoskeletons (Turner and Adler, 1998). Substantial variation is seen in the severity of the inhibition of the putative refinement and in the number of hairs produced by individual cells (Fig. 2). The relationship between the *fz* pathway and the refinement genes appears to be complex. Double mutants between genes such as *in* or *mwh* and several of the refinement genes give additive phenotypes suggesting that these genes function in parallel pathways (Cong et al., 2001). However, as noted below, there is evidence that the *fz* pathway regulates at least some of the refinement genes. It is presumed that mutations that result in multiple hairs of normal polarity do not disrupt the accumulation of distal and proximal planar polarity protein complexes. This subject has not been intensively studied but in limited experiments we have found this to be the case.

The activation of the cytoskeleton to form a hair is likely to involve RhoA, Rho Kinase, and the cellular myosins encoded by *crinkled* (a myosin VII), *zipper* (a myosin II) and *spaghetti squash* (which encodes an essential light chain) (Strutt et al., 1997; Winter et al., 2001). As noted above loss of function mutations in these genes (which I will refer to as *rok* pathway) result in multiple hairs forming at the normal distal side of the cell. This indicates that the function of these proteins are not required downstream of the *fz*-like genes for specifying the location of hair formation. Gene dosage interactions have argued that these genes are downstream of the *fz*-like genes and there is strong evidence that *fz/dsh* activate *Drok* leading to the phosphorylation of MRLC and the activation of myosin II (Winter et al., 2001). Opposite gene dosage interactions were seen with *ck*. This activation of the cytoskeleton by the *fz* pathway appears to be independent of the *inturned*-like genes as gene dosage interactions equivalent to those seen between *rok* pathway and *fz*-like genes were not seen between *inturned*-like and *rok* pathway genes. It seems likely that the *fz* pathway branches before interacting with either the *inturned* or *rok* pathways. The activation of *Drok* by *fz/dsh* might correspond to the "late *fz* gain of function" (Krasnow and Adler, 1994).

Acknowledgments

The work in my laboratory is supported by the NIGMS.

References

Adler, P., Taylor, J., Charlton, J. 2000. The domineering non-autonomy of frizzled and van Gogh clones in the *Drosophila* wing is a consequence of a disruption in local signaling. Mech. Dev. 96, 197–207.

Adler, P.N. 2002. Planar signaling and morphogenesis in *Drosophila*. Dev. Cell 2, 525–535.

Adler, P.N., Charlton, J., Jones, K.H., Liu, J. 1994. The cold-sensitive period for frizzled in the development of wing hair polarity ends prior to the start of hair morphogenesis. Mech. Dev 46, 101–107.

Adler, P.N., Charlton, J., Liu, J. 1998. Mutations in the cadherin superfamily member gene dachsous cause a tissue polarity phenotype by altering frizzled signaling. Development 125, 959–968.

Adler, P.N., Krasnow, R.E., Liu, J. 1997. Tissue polarity points from cells that have higher Frizzled levels towards cells that have lower Frizzled levels. Cur. Biol. 7, 940–949.

Axelrod, J.D. 2001. Unipolar membrane association of Disheveled mediates Frizzled planar cell polarity signaling. Genes. Dev. 15, 1182–1187.

Bastock, R., Strutt, H., Strutt, D. 2003. Strabismus is asymmetrically localised and binds to Prickle and Disheveled during *Drosophila* planar polarity patterning. Development 130, 3007–3014.

Carreira-Barbosa, F., Concha, M.L., Takeuchi, M., Ueno, N., Wilson, S.W., Tada, M. 2003. Prickle 1 regulates cell movements during gastrulation and neuronal migration in zebrafish. Development 130, 4037–4046.

Chae, J., Kim, M.J., Goo, J.H., Collier, S., Gubb, D., Charlton, J., Adler, P.N., Park, W.J. 1999. The *Drosophila* tissue polarity gene *starry night* encodes a member of the protocadherin family. Development 126, 5421–5429.

Clark, H.F., Brentrup, D., Schneitz, K., Bieber, A., Goodman, C., Noll, M. 1995. Dachsous encodes a member of the cadherin superfamily that controls imaginal disc morphogenesis in *Drosophila*. Genes. Dev. 9, 1530–1542.

Collier, S., Gubb, D. 1997. *Drosophila* tissue polarity requires the cell-autonomous activity of the fuzzy gene, which encodes a novel transmembrane protein. Development 124, 4029–4037.

Cong, J., Geng, W., He, B., Liu, J., Charlton, J., Adler, P. 2001. The furry gene of *Drosophila* is important for maintaining the integrity of cellular extensions during morphogenesis. Development 128, 2793–2802.

Curtin, J.A., Quint, E., Tsipouri, V., Arkell, R.M., Cattanach, B., Copp, A.J., Henderson, D.J., Spurr, N., Stanier, P., Fisher, E.M., et al. 2003. Mutation of Celsr1 disrupts planar polarity of inner ear hair cells and causes severe neural tube defects in the mouse. Curr. Biol. 13, 1129–1133.

Darken, R.S., Scola, A.M., Rakeman, A.S., Das, G., Mlodzik, M., Wilson, P.A. 2002. The planar polarity gene strabismus regulates convergent extension movements in Xenopus. EMBO J. 21, 976–985.

Das, G., Reynolds-Kenneally, J., Mlodzik, M. 2002. The Atypical Cadherin Flamingo Links Frizzled and Notch Signaling in Planar Polarity Establishment in the *Drosophila* eye. Dev. Cell 2, 655–666.

Fanto, M., Clayton, L., Meredith, J., Hardiman, K., Charroux, B., Kerridge, S., McNeill, H. 2003. The tumor-suppressor and cell adhesion molecule Fat controls planar polarity via physical interactions with Atrophin, a transcriptional co-repressor. Development 130, 763–774.

Feiguin, F., Hannus, M., Mlodzik, M., Eaton, S. 2001. The ankyrin repeat protein Diego mediates Frizzled-dependent planar polarization. Dev. Cell 1, 93–101.

Goto, T., Keller, R. 2002. The planar cell polarity gene strabismus regulates convergence and extension and neural fold closure in Xenopus. Dev. Biol. 247, 165–181.

Gubb, D., Garcia-Bellido, A. 1982. A genetic analysis of the determination of cuticular polarity during development in *Drosophila* melanogaster. J. Embryol. Exp. Morphol. 68, 37–57.

Gubb, D., Green, C., Huen, D., Coulson, D., Johnson, G., Tree, D., Collier, S., Roote, J. 1999. The balance between isoforms of the prickle LIM domain protein is critical for planar polarity in *Drosophila* imaginal discs. Genes. Dev. 13, 2315–2327.

He, B., Adler, P.N. 2002. The frizzled pathway regulates the development of arista laterals. BMC Dev. Biol. 2, 7.

Heisenberg, C.P., Tada, M., Rauch, G.J., Saude, L., Concha, M.L., Geisler, R., Stemple, D.L., Smith, J.C., Wilson, S.W. 2000. Silberblick/Wnt11 mediates convergent extension movements during zebrafish gastrulation. Nature 405, 76–81.

Kiehart, D.P., Montague, R.A., Roote, J., Ashburner, M. 1999. Evidence that crinkled, mutations in which cause numerous defects in *Drosophila* morphogenesis, encodes a myosin VII. InA. Dros. Res. Conf. 40 (ed., pp. 295C).

Klingensmith, J., Nusse, R., Perrimon, N. 1994. The *Drosophila* segment polarity gene disheveled encodes a novel protein required for response to the wingless signal. Genes. Dev 8, 118–130.

Krasnow, R.E., Adler, P.N. 1994. A single frizzled protein has a dual function in tissue polarity. Development 120, 1883–1893.

Lawrence, P.A. 1966. Gradients in the insect segment: The orientation of hairs in the milkweed bug *Oncopeltus fasciatus*. J. Exp. Biol. 44, 607–620.

Lawrence, P.A., Casal, J., Struhl, G. 2002. Towards a model of planar polarity and pattern in the *Drosophila* abdomen. Development 129, 2749–2760.

Lee, H., Adler, P.N. 2002. *inturned* and *fuzzy* are essential for the function of the *frizzled* pathway in the wing. Genetics 160, 1535–1547.

Lu, B., Usui, T., Uemura, T., Jan, L., Jan, Y.N. 1999. Flamingo controls the planar polarity of sensory bristles and asymmetric division of sensory organ precursors in *Drosophila*. Cur. Biol. 9, 1247–1250.

Ma, D., Yang, C.H., McNeil, H., Simon, M.A., Axelrod, J.D. 2003. Fidelity in planar cell polarity signaling. Nature 421, 543–547.

Mahoney, P.A., Weber, U., Onofrechuk, P., Biessmann, H., Bryant, P.J., Goodman, C.S. 1991. The fat tumor suppressor gene in *Drosophila* encodes a novel member of the cadherin gene superfamily. Cell 67, 853–868.

Mlodzik, M. 2002. Planar cell polarization: Do the same mechanisms regulate *Drosophila* tissue polarity and vertebrate gastrulation? Trends Genet. 18, 564–571.

Montcouquiol, M., Rachel, R.A., Lanford, P.J., Copeland, N.G., Jenkins, N.A., Kelley, M.W. 2003. Identification of *Vangl2* and *Scrb1* as planar polarity genes in mammals. Nature 423, 173–177.

Park, W.J., Liu, J., Sharp, E.J., Adler, P.N. 1996. The *Drosophila* tissue polarity gene inturned acts cell autonomously and encodes a novel protein. Development 122, 961–969.

Shimada, Y., Usui, T., Yanagawa, S., Takeichi, M., Uemura, T. 2001. Asymmetric colocalization of Flamingo, a seven-pass transmembrane cadherin, and Disheveled in planar cell polarization. Cur. Biol. 11, 859–863.

Strutt, D. 2001a. Planar polarity: Getting ready to ROCK. Current Biology 11, R506–R509.

Strutt, D. 2002. The asymmetric subcellular localisation of components of the planar polarity pathway. Seminars in Cell and Developmental Biology 13, 225–231.

Strutt, D.I. 2001. Asymmetric localization of frizzled and the establishment of cell polarity in the *Drosophila* wing. Mol. Cell 7, 367–375.

Strutt, D.I., Johnson, R., Cooper, K., Bray, S. 2002. Asymmetric localisation of Frizzled and the determination of Notch-dependent cell fate in the *Drosophila* eye. Cur. Biol. 12, 813–824.

Strutt, D.I., Weber, U., Mlodzik, M. 1997. The role of RhoA in tissue polarity and Frizzled signaling. Nature 387, 292–295.

Strutt, H., Strutt, D.I. 2002. Nonautonomous planar polarity patterning in *Drosophila*: *Disheveled* independent functions of *frizzled*. Dev. Cell. 3, 851–863.

Taylor, J., Abramova, N., Charlton, J., Adler, P.N. 1998. Van Gogh: A new *Drosophila* tissue polarity gene. Genetics 150, 199–210.

Theisen, H., Purcell, J., Bennett, M., Kansagara, D., Syed, A., Marsh, J.L. 1994. Disheveled is required during wingless signaling to establish both cell polarity and cell identity. Development 120, 347–360.

Tomlinson, A., Struhl, G. 1999. Decoding vectorial information from a gradient: sequential roles of the receptors Frizzled and Notch in establishing planar polarity in the *Drosophila* eye. Development 126, 5725–5738.

Tree, D.R.P., Shulman, J.M., Rousset, R., Scott, M.P., Gubb, D., Axelrod, J.D. 2002. Prickle Mediates Feedback Amplification to Generate Asymmetric Planar Cell Polarity Signaling. Cell 109, 1–11.

Turner, C.M., Adler, P.N. 1995. Morphogenesis of *Drosophila* pupal wings *in vitro*. Mech. Dev. 52, 247–255.

Turner, C.M., Adler, P.N. 1998. Distinct roles for the actin and microtubule cytoskeletons in the morphogenesis of epidermal hairs during wing development in *Drosophila*. Mech. Dev. 70, 181–192.

Usui, T., Shima, Y., Shimada, Y., Hirano, S., Burgess, R.W., Schwarz, T.L., Takeichi, M., Uemura, T. 1999. Flamingo, a seven-pass transmembrane cadherin, regulates planar cell polarity under the control of Frizzled. Cell 98, 585–595.

Veeman, M.T., Axelrod, J.D., Moon, R.T. 2003a. A Second Canon: Function of Mechanisms of b-Catenin-Independent Wnt Signaling. Dev. Cell 5, 367–377.

Veeman, M.T., Slusarski, D.C., Kaykas, A., Louie, S.H., Moon, R.T. 2003b. Zebrafish prickle, a modulator of noncanonical wnt/fz signaling, regulates gastrulation movements. Cur. Biol. 13, 680–685.

Vinson, C.R., Adler, P.N. 1987. Directional non-cell autonomy and the transmission of polarity information by the frizzled gene of *Drosophila*. Nature 329, 549–551.

Vinson, C.R., Conover, S., Adler, P.N. 1989. A *Drosophila* tissue polarity locus encodes a protein containing seven potential transmembrane domains. Nature 338, 263–264.

Winter, C.G., Wang, B., Ballew, A., Royou, A., Karess, R., Axelrod, J.D., Luo, L. 2001. *Drosophila* rho-associated kinase (Drok) links frizzled-mediated planar cell polarity signaling to the actin cytoskeleton. Cell 105, 81–91.

Wolff, T., Rubin, G. 1998. Strabismus, a novel gene that regulates tissue polarity and cell fate decisions in *Drosophila*. Development 125, 1149–1159.

Wong, L.L., Adler, P.N. 1993. Tissue polarity genes of *Drosophila* regulate the subcellular location for prehair initiation in pupal wing cells. J. Cell Biol. 123, 209–221.

Yang, C.H., Axelrod, J.D., Simon, M.A. 2002. Regulation of Frizzled by Fat-like Cadherins during planar polarity signaling in the *Drosophila* Compound Eye. Cell 108, 675–688.

Zeidler, M.P., Perrimon, N., Strutt, D.I. 2000. Multiple roles for four-jointed in planar polarity and limb patterning. Dev. Biol. 228, 181–196.

Zheng, L., Zhang, J., Carthew, R.W. 1995. Frizzled regulates mirror-symmetric pattern formation in the *Drosophila* eye. Development 121, 3045–3055.

Planar cell polarity in the *Drosophila* eye: Cell fate and organization

Marek Mlodzik

*Mount Sinai School of Medicine, Department of Molecular, Cell and Developmental Biology,
New York, New York 10029*

Contents

1. Introduction

1.1. General comments

In multicellular organisms, most tissues derived from epithelial cell sheets form highly organized structures that are not only polarized in the apical-basolateral axis but also display a polarization within the plane of the epithelium (Eaton, 1997). In

Advances in Developmental Biology
Volume 14 ISSN 1574-3349
DOI: 10.1016/S1574-3349(04)14002-7

Drosophila, this is evident in all tissues derived from imaginal disc epithelia, including the eye (Adler, 1992, 2002; Mlodzik, 2002). The function of many organs or tissues requires this additional axis of polarity within the plane of the epithelium, reflected in a uniform polarity of single cells or multicellular units. This type of polarization of cells is usually referred to as epithelial planar cell polarity (PCP), or in *Drosophila*, often, as tissue polarity (Adler, 1992, 2002; Mlodzik, 2002). The general aspects of PCP establishment and the organ- or tissue-specific features are discussed in many of the other chapters in this book. This chapter focuses on the specific features of PCP establishment, signaling, and cellular read-out in the *Drosophila* eye (Blair, 1999; Mlodzik, 1999; Reifegerste and Moses, 1999; Strutt and Strutt, 1999), one of several *Drosophila* tissues with specific advantages for the genetic dissection and analysis of the process. The *Drosophila* eye is not the only neuroepithelium with well established PCP features and with experimental dissection of its establishment. Other neuro-epithelia with PCP features include the mammalian inner ear epithelium, the cochlea, where the stereocilial bundles are aligned for normal sensitivity to sound (Curtin et al., 2003; Dabdoub et al., 2003; Montcouquiol et al., 2003) (see also the chapter by Dabdoub et al. in this book), or the orientation of the sensory bristles on the *Drosophila* thorax (Although PCP establishment on the *Drosophila* thorax is not covered in this book, there are several excellent articles and reviews covering this tissue; please see Bellaiche et al., 2004; Gho and Schweisguth, 1998).

The presence of PCP features in the *Drosophila* eye (Fig. 1), and in compound insect eyes in general, were first noticed a long time ago, dating back almost 100 years (e.g., Dietrich, 1909; Lawrence and Shelton, 1975). The study of the establishment of PCP in the *Drosophila* retina serves as a paradigm for PCP establishment in multicellular units (Blair, 1999; Mlodzik, 1999; Strutt and Strutt, 1999). The purpose of this chapter is to review the molecular aspects of PCP signaling in the fly eye, and compare it to PCP signaling in other fly tissues and in general, to discuss the eye-specific and common read-outs resulting from the interpretation of the initial PCP signaling.

1.2. Why is the fly retina polarized?

Like most (if not all) adult structures, the *Drosophila* retina displays a characteristic PCP arrangement (Fig. 1; Dietrich, 1909; Blair, 1999; Mlodzik, 1999; Strutt and Strutt, 1999; Adler, 2002; Mlodzik, 2002; Strutt, 2003). However, the retina is a complex neural sensory organ structure, where multiple cells form each ommatidial unit (also called 'facet'), and each is composed of eight photoreceptor neurons and twelve accessory cells (see Wolff and Ready, 1993 for details); and thus it is not a "simple" epithelial tissue, and the polarization of single cells is only meaningful within the context of each unit. What does the PCP arrangement mean in the retina and why is it important?

As the function of the visual system is to receive and transmit information to the brain to form images, precise retinotopic projections of retinal photoreceptors onto the optic lobes in the brain are an important prerequisite. The *Drosophila* eye, like all insect eyes, is a compound eye, containing several hundred ommatidia, each representing a separate unit. Each point in space is perceived by one ommatidium.

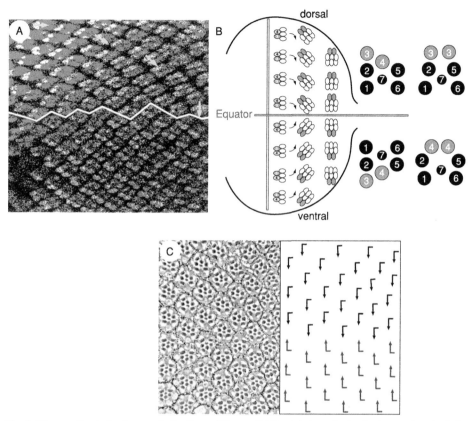

Fig. 1. Planar cell polarity features in the *Drosophila* eye. (A) Partial view of a developing eye imaginal disc demonstrating the regularity of polarity establishment. Anterior is left and dorsal up in this and all subsequent figures. Ommatidial clusters are marked with anti-Elav (green; labeling all photoreceptors) and *svp-lacZ* (magenta; *svp* is expressed initially in R3/R4—see left side of panel—and later also in R1/R6 at weaker levels). The morphogenetic furrow is on the left side adjacent to the field shown. The orientation and degree of rotation of some dorsal ommatidial preclusters is highlighted with yellow arrows; white line marks the equator. (B) Schematic drawing of third instar larval eye imaginal disc, with the morphogenetic furrow (M; yellow) and the D/V midline (the equator; gray) indicated. Initially, ommatidial preclusters are symmetrical and organized in the A/P axis. Subsequently they rotate 90° with respect to the equator; at the end of this process chirality is established by the positions of R3 and R4. Right side: schematic presentation of chiral organization of dorsal and ventral adult ommatidia; in addition to the chiral forms, symmetrical clusters with R3/R3 or R4/R4 cell pairs (as often found in PCP mutant tissue) are also shown. R3 cells are highlighted in green and R4s in magenta. (C) Tangential section of a wild-type adult eye (left panel) with the respective schematic presentation (right panel). The dorsal and ventral ommatidial arrangements are represented by black and red arrows, respectively. Note the very regular ommatidial arrangement and the line of mirror image symmetry between the dorsal and ventral halves. (See Color Insert.)

As therefore only the composite input of several ommatidia can create an image, the alignment of a single ommatidium needs to be very precise with respect to its neighbors and the whole eye field. Thus, a whole aspect of *Drosophila* eye development and patterning relates to the precise organization of the ommatidia with respect to their neighboring ommatidia and their position within the eye. Again, the establishment of this very precise arrangement is mediated by the PCP genes (Dietrich, 1909; Blair, 1999; Mlodzik, 1999; Strutt and Strutt, 1999; Adler, 2002; Mlodzik, 2002; Strutt, 2003), and it is not only necessary for the proper arrangement of single ommatidia within the eye, but also for the proper arrangement of photoreceptors within each ommatidium itself. Both aspects are critical for the correct innervation and neuronal connectivity in the optic lobes (Clandinin and Zipursky, 2000, 2002) and thus for image formation and accurate vision in general.

2. The arrangement of the ommatidia within the plane of the retina

2.1. Establishment of ommatidial polarity during development

The *Drosophila* eye is polarized in a spectacular way, and the precise orientation of single ommatidia (or facets) is reflected in their mirror image arrangement of opposite chiral forms across the dorso-ventral (D/V) midline and their precise alignment in the antero–posterior (A/P) axis (Wolff and Ready, 1991, 1993). Thus all ommatidia are arranged with respect to both, the A/P and the D/V axes (Fig. 1; see also reviews by Blair, 1999; Mlodzik, 1999; Reifegerste and Moses, 1999; Strutt and Strutt, 1999). The antero–posterior arrangement is established by the direction of the progression of the morphogenetic furrow (MF) and is the first visible orientation of the developing ommatidial preclusters (reviewed in Heberlein and Moses, 1995; Treisman and Heberlein, 1998). Subsequently, the D/V alignment is generated in response to a polarizing signal that organizes the ommatidia around the dorsoventral midline, the so-called equator. The critical signaling events that directly govern the D/V organization of the single ommatidia occur posterior to the MF, before or at the stage of the 5-cell precluster, and are mediated by the *frizzled*–PCP pathway and associated genes (Zheng et al., 1995; Strutt et al., 1997; Tomlinson et al., 1997; Wehrli and Tomlinson, 1998).

How is this complex and precise orientation established? As the MF advances anteriorly, ommatidial preclusters emerge posterior to it, having a single axis of symmetry and facing in the same direction. Within the following 6–8 hours the clusters rotate by 45° away from the A/P axis (Fig. 1A,B). They maintain this angle for about 8 hours, before rotating a further 45°, bringing them to a final 90° from their original position in the A/P axis. Subsequently, the ommatidial clusters remain in this orientation (for additional reading see Wolff and Ready, 1993). Ommatidia in the dorsal and ventral halves of the eye imaginal disc rotate in opposite directions, thus creating a line of mirror-image symmetry, generally called the equator as it is running along the D/V midline. Although the different chiralities of the dorsal and ventral clusters only become morphologically evident in mature ommatidia, where the R3 and R4 photoreceptors are asymmetrically positioned at the tip of the ommatidial trapezoids (Fig. 1B,C), molecular markers for the early R3 and R4 photoreceptors show that chirality establishment actually precedes ommatidial rotation (see below). It is thus presumed that the direction of rotation, directly follows from the establishment of cell fate governing the chirality decision (reviewed in Mlodzik, 1999; Strutt and Strutt, 1999).

Initially, as the developing ommatidial precluster emerges from the furrow, the R3/R4 precursor pair is symmetrically arranged in the eye imaginal discs. At this stage the R3 precursor cell is closer to the D/V midline/equator than the R4 precursor (see Fig. 1B for a schematic presentation). The first morphological sign of asymmetry is the beginning of ommatidial rotation. This is accompanied and even preceded by asymmetric expression of several molecular markers, highlighting the loss of symmetry. Thus, during the first 4–5 rows posterior to the furrow, the basis for the formation of opposite chiral forms in the ventral and dorsal halves is established, as the R3 and R4 precursors become specified as distinct cells. This is later reflected in their asymmetric positioning within the mature ommatidial cluster, forming the chiral trapezoids (Fig. 1B,C) (Dietrich, 1909; Tomlinson and Ready, 1987; Tomlinson, 1988; Wolff and Ready, 1991,1993; Mlodzik, 1999; Reifegerste and Moses, 1999; Strutt and Strutt, 1999).

The features of PCP establishment mentioned above have suggested that the polarizing signal during eye development originates at the D/V midline, the equator. This factor is often referred to as factor X (Wehrli and Tomlinson, 1995; Wehrli and Tomlinson, 1998). How the D/V midline (equator) is established, which is also critical for the definition of the point where the furrow initiates, is discussed in detail in the chapter by Kwang Choi in this book (see also references therein). In brief, recent work has implicated several signaling molecules and pathways in setting up the dorso-ventral midline in the second larval instar eye disc, which serves later in the third instar larvae as the equator. These include *pannier*, the Wg pathway, the homeo domain genes of the *Iroquois* complex, the Notch pathway, and the *hopscotch/* JAK-STAT pathway. Ultimately, the interplay of the initial signals leads to the expression *fringe* in the ventral half of the disc, which mediates Notch activation at the boundary of *fringe* expressing and non-expressing cells (see the chapter by Singh et al. in this book) (also reviewed in Strutt and Strutt, 1999). The 'extracellular regulation' of PCP signaling and the establishment of the expression patterns of the potential regulators of this aspect are discussed in detail in the chapters by Strutt & Strutt and Choi in this book (see also references and description therein).

2.2. *The role of the R3 and R4 photoreceptors in ommatidial polarization*

Genetic analysis of mutants affecting PCP establishment has indicated that the R3/R4 photoreceptor precursor pair is critical for the establishment of ommatidial polarity. The first evidence for this requirement came from the elegant clonal analysis with *frizzled* (Zheng et al., 1995). The initial step in the cascade of events required for the correct and distinct determination of R3 and R4 is the specification of the R3/R4 pair as special photoreceptor class (to allow them subsequently to properly interpret the PCP signals). The correct R3/R4 subtype specification is mediated by (at least) two transcription factors, the nuclear receptor Seven-up (Fanto et al., 1998) and the Zinc-finger proteins of the Spalt complex (Domingos et al., 2004). Following this subtype specification, the correct and distinct cell fate of the R3 and R4 photoreceptors determines the polarity, direction of rotation, and the resulting

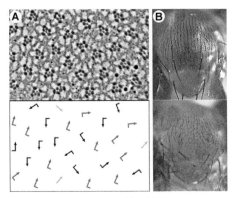

Fig. 2. A typical retinal PCP mutant phenotype. (A). Homozygous mutant *dsh*–eye tissue and its schematic presentation in lower panel. Arrows are drawn as in Fig. 1 with black and red arrows representing dorsal and ventral ommatidial chirality, respectively, and symmetrical R3/R3 or R4/R4 ommatidia are represented by green arrows. The equatorial arrangement is lost with a random arrangement of both chiralities (black and red arrows), and the presence of several symmetrical clusters of the R3/R3 or R4/R4 types (green arrows), compare to Fig. 1C for regular wild–type arrangement. (B) PCP arrangement of the sensory touch bristles on the dorsal notum for comparison (upper panel: *wild type*; lower panel: *dsh*–). Note that in both tissues the orientation of the respective neural sensory units is random relative to the body axes. (See Color Insert.)

chirality of the ommatidium. Based on the positions of the R3 and R4 precursors in the 5-cell precluster, R3 being on the equatorial side and R4 on the polar side of the cluster, it is thought that the polarizing signal is received first, or at higher levels, in the R3 precursor as it is closer to the presumed source (of the signal). This would predispose the R3 precursor to be specified as the R3 cell, and in turn would then lead to the distinct specification of the neighboring R4. The cell fate specification is then followed by the direction of rotation.

In mutants affecting polarity in the eye (see below) both the R3/R4 specification (and associated chirality) and the direction and degree of rotation become largely randomized (Gubb, 1993; Theisen et al., 1994; Zheng et al., 1995; Mlodzik, 1999; Strutt and Strutt, 1999). An example of a retinal PCP mutant phenotype is shown in Fig. 2A. Strikingly, in many PCP mutants some ommatidial clusters remain symmetrical, giving rise to either V- or U-shaped adult ommatidia with non-chiral R3/R3 or R4/R4 photoreceptor pair arrangements, confirming the importance of cell fate specification within the R3/R4 pair in this context (see Figs. 1B and 2).

3. Genetic control of PCP establishment in the retina

3.1. The planar cell polarity genes

Many of the components required for PCP establishment in the retina have been identified in genetic screens in other tissues, and are thus generally required for PCP in most (if not all) *Drosophila* tissues (e.g., wing, legs, abdomen, and dorsal thorax). These are commonly referred to as "primary polarity genes" (Adler, 1992; Gubb, 1993; Adler, 2002; Mlodzik, 2002; Strutt, 2003). They comprise the following group: *frizzled* (*fz*) (Zheng et al., 1995; Vinson and Adler, 1987; Vinson et al., 1989), *disheveled* (*dsh*) (Klingensmith et al., 1994; Theisen et al., 1994; Boutros and Mlodzik, 1999), *prickle-spiny legs* (*pk-sple*) (Gubb et al., 1999; Mlodzik, 2000; Tree et al., 2002), *diego* (*dgo*) (Feiguin et al., 2001; Das et al., 2004), *rhoA* (Strutt et al., 1997), *misshapen* (*msn*) (Paricio et al., 1999), *strabismus*, also called *Van Gogh* (-*stbm*/*Vang*) (Taylor et al., 1998; Wolff and Rubin, 1998) and *flamingo*, also called *starry night* (*fmi*/*stan*) (Chae et al., 1999; Usui et al., 1999; Das et al., 2002). Whereas the primary PCP genes affect both chirality and direction (and degree) of rotation, the eye-specific polarity genes appear only to affect the rotation of the ommatidia. They do not affect the R3/R4 cell fate decision and thus the chirality generation (Choi and Benzer, 1994; Brown and Freeman, 2003; Gaengel and Mlodzik, 2003; Strutt and Strutt, 2003) (see section 5 of this chapter).

The most prominent tissue polarity gene is *frizzled* (*fz*). Elegant work by Paul Adler and colleagues has led to a detailed understanding of its phenotypic and molecular features (Vinson and Adler, 1987; Vinson et al., 1989; Adler et al., 1990; Krasnow and Adler, 1994; Wang et al., 1994; Jones et al., 1996) (see also the chapter by Paul Adler in this book). This was followed by work in Richard Carthew's lab showing that *fz* is required in R3 in the eye (Zheng et al., 1995). Among the PCP genes, *fz* displays a unique feature. Whereas most other PCP genes are required cell autonomously in the mutant cells, *fz⁻* tissue appears also to affect neighboring wild-type cells on the distal (wing) or polar (eye) side of the mutant clone (Vinson and Adler, 1987; Zheng et al., 1995) (with the other exception being *stbm*/*Vang* (Taylor et al., 1998) (see also the chapter by Paul Adler in this book). One of the interpretations of this observation has been that (in addition to the cell-autonomous signaling requirement of *fz*) the "signal" cannot travel through a *fz* patch of cells, implying that *fz* is not only necessary for reading the signal but also for its propagation. Nevertheless, there are several other potential explanations that are equally more likely (Tree et al., 2002; Lawrence et al., 2004), and as the molecular nature of *fz* activation and its potential ligand remain elusive (see below) this question cannot be answered as of now.

Molecular cloning of *fz* indicated that it might function as a receptor: Its primary sequence predicted a seven-pass transmembrane-receptor-like molecule [Vinson, 1989, #30]. Subsequently, the *frizzled* family of transmembrane receptors has indeed been shown to act as receptors, specifically as the receptors for the Wnt family of secreted growth factors (Bhanot et al., 1996). In several developmental contexts (outside of PCP signaling) *fz* can act as the receptor for Wg itself (this function is, however, redundant with *fz2* (Bhat, 1998; Kennerdell and Carthew, 1998; Bhanot et al., 1999; Chen and Struhl, 1999; Mueller et al., 1999)). Again, as the PCP-specific *fz* ligand has not yet been identified, it appears now unlikely that it is another member of the Wnt family, and thus it remains completely open as to what this elusive PCP ligand (if any) might be.

Of the other PCP genes, *disheveled* (*dsh*) is the best characterized (Klingensmith et al., 1994; Theisen et al., 1994; Krasnow et al., 1995; Axelrod et al., 1998; Boutros et al., 1998; Boutros and Mlodzik, 1999). In analogy to canonical Wnt/*fz* signaling through β-Catenin, genetic epistasis analysis has placed *dsh* downstream of *fz* in PCP signaling (Krasnow et al., 1995; Strutt et al., 1997). The *dsh* gene encodes a 70kd cytoplasmic protein with no known biochemical function (Klingensmith et al., 1994; Krasnow et al., 1995; Yanagawa et al., 1995). Homologues of *dsh* have been identified in many organisms, ranging from nematodes to humans (reviewed in Boutros and Mlodzik, 1999). All *dsh* proteins identified share three highly conserved domains: a DIX domain, a central PDZ domain, and a C-terminal DEP domain (reviewed in Boutros and Mlodzik, 1999). All three domains have been implicated as protein–protein interaction modules (and shown to interact with several other proteins), and thus *dsh* appears to serve as an adapter molecule. Recently, *dsh* has been molecularly linked to *fz* through an interaction between the *dsh*–PDZ domain and amino acids in the cytoplasmic tails of *fz* receptors (Wong et al., 2003). This appears common to both signaling pathways mediated by *fz–dsh* (see below). It remains unclear if there are pathway-specific *fz–dsh* interactions.

    In the PCP context in the fly retina and thus R3/R4 specification, *fz* and *dsh* are both required to generate a bias within the early symmetrical R3/R4 pair (Zheng et al., 1995; Strutt et al., 1997). More specifically, *fz* and *dsh* specify the R3 fate (Fanto and Mlodzik, 1999; Tomlinson and Struhl, 1999). The particular roles of the other primary PCP genes in regulating *fz–dsh* activity are discussed below.

3.2. *A* frizzled *mediated PCP signaling pathway*

    Several primary PCP genes and other signaling components interact genetically with *fz* and *dsh*. Depending on the tissue, the effector pathways downstream of *fz–dsh* show variability due to the differences in the tissue-specific responses to the PCP pathway. In the wing, for example, the main read-out is the localization of the actin assembly "machinery" to one (distal) area within each cell (reviewed in Adler, 2002; Mlodzik, 2002; Strutt, 2003). In contrast, in the eye, an important read-out is the cell fate specification of the R3 and R4 photoreceptors as distinct cells, and thus a nuclear read-out and transcriptional event are a key aspect downstream of *fz–dsh* in this tissue (reviewed in Adler, 2002; Mlodzik, 1999; Strutt and Strutt, 1999).

    The *fz–dsh*–PCP pathway acting in the *Drosophila* eye, and probably also in other multicellular context involving cell fate specification, has been assembled through a combination of genetic and biochemical studies. Downstream of *fz* and *dsh*, the pathway consists of the small GTPase of the *rhoA* family and the STE20-like kinase *misshapen* (*msn*) (Strutt et al., 1997; Boutros et al., 1998; Paricio et al., 1999; Fanto et al., 2000). Both *rhoA* and *msn* are also required for the generation of PCP in most (if not all) tissues analyzed (Strutt et al., 1997; Paricio et al., 1999). In addition, genetic interactions, cell culture experiments, and phenotypic analyses also indicate that a JNK-type MAPK module and the AP-1 transcription factor act downstream of *dsh* and *msn* in eye PCP establishment (Boutros et al., 1998; Paricio et al., 1999; Weber et al., 2000). All the factors acting downstream of *dsh* are not appreciably involved in canonical Wnt/β-Catenin signaling (Axelrod et al., 1998; Boutros et al., 1998), and thus a signaling cascade that is distinct from the canonical Wg/Wnt pathway has emerged for PCP establishment and is generally referred to as the *fz*–PCP pathway.

The small GTPases of the Rac subfamily (represented by *Drac1* and *Drac2* in *Drosophila*) appear also to be involved in the pathway regulating PCP in the retina (Fanto et al., 2000) and also the wing (Eaton et al., 1996). Based on dominant negative and gain-of-function studies and supported by genetic interactions with the respective *Drac1* and *Drac2* deficiencies, Rac is thought to mediate the activation of *msn* and/or the JNK-cascade downstream of *dsh* (Boutros et al., 1998; Fanto et al., 2000). These observations indicate that in the eye, *fz* signals to the nucleus via a *dsh/*Rac/*msn* pathway that leads to the activation of JNK- (and p38-) type kinase cascades and Jun (Fanto et al., 2000; Weber et al., 2000) (summarized in Fig. 3). Strikingly, the contribution of the Rac GTPase subfamily (as part of the *rho* class) must be redundant with either *rhoA* itself or *cdc42* as *Rac1,2-/-*tissue appears to develop normal PCP features both in the eye and in the wing (Hakeda-Suzuki et al., 2002), although the use of dominant negative and activated isoforms of Rac clearly indicates a requirement in both tissues (Eaton et al., 1996; Fanto et al., 2000). Our unpublished observations suggest indeed that the Rac GTPases act redundantly with both Cdc42 and *rhoA*, but further experiments are needed to dissect this problem in sufficient detail (Silvia Munoz, MM, and Nuria Paricio, unpublished).

The specific involvement and requirements of the kinases acting downstream of *msn* as determined by genetic interactions and biochemical studies remains also unclear. Both *Drosophila dsh* and its human homologs act as potent activators of JNK signaling (Boutros et al., 1998; Li et al., 1999). However, although mutations in components of the JNK-cascade dominantly suppress gain-of-function genotypes of *fz, dsh,* or *msn* they do not show strong PCP phenotypes in simple loss-of-function analyses, indicating that there is again redundancy at this level in the cascade (Boutros et al., 1998; Paricio et al., 1999; Fanto et al., 2000; Weber et al., 2000). Genetic experiments suggest that the observed redundancy of the JNK module is due to the action of the related p38-type MAPKinases (Paricio et al., 1999; Weber et al., 2000). As components of either kinase cascade are able to cross-phosphorylate the respective recipients in either module, this possibility is supported by existing biochemical analysis in several vertebrate cell culture systems.

Despite the identification of the new components of *fz*–PCP signaling, their molecular links are not resolved. Although *fz* recruits *dsh* to the membrane by direct binding (Wong et al., 2003), the mechanism for how this is achieved or how *dsh* activates further downstream PCP effectors like *rhoA*, Rac, and/or *msn*/STE20 remains unclear.

3.3. Regulation of frizzled activity: upstream-extracellular factors

Despite the increasing knowledge about the role of *fz* and its associated pathway in PCP generation and the insights of how the D/V-midline, the equator, is established (see chapters by K. Choi and Strutt & Strutt in this book), the ligand for *fz* in the PCP context is still unknown. Based on the fact that *fz* family receptors generally bind

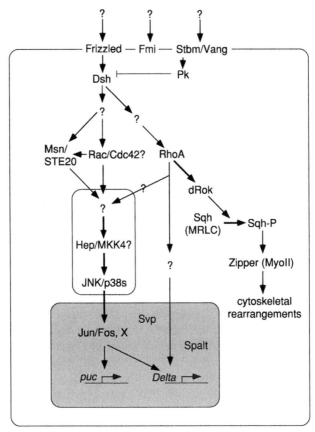

Fig. 3. Schematic model of the *frizzled*–PCP signaling pathway. The PCP effector pathway downstream of *fz–dsh* is distinct from the canonical Wnt pathway and complex. Depending on the tissue, a nuclear, and/ or a cytoskeletal response is activated. In the eye, it appears that both branches are important for either cell fate specification of R3/R4 (nuclear) or ommatidial rotation (cytoskeletal-dRok branch). Several of the other PCP genes regulate the activity of *fz–dsh* at the cell membrane. See also Figure 5 for more details. In the R3/R4 photoreceptor cells several transcription factors (Svp, Spalt, and others) are needed for correct cell fate interpretation that is a prerequisite for the normal function of the *fz–dsh*–PCP pathway. See text for details.

Wnt ligands (Bhanot et al., 1996), a candidate for the planar polarity *fz* ligand should be a Wnt gene. Nevertheless, so far none of the Wnt genes analyzed in sufficient detail appears to be the right candidate. Thus, one might speculate that the 'search' has been going in the wrong direction and that the *fz*–PCP ligand is not a Wnt. This could be envisioned as either a 'heterologous' ligand binding directly to *fz* (e.g., the example of mammalian Norrin as a high affinity ligand for *fz*4 comes to mind in this regard (Xu et al., 2004)), or that *fz* is part of a receptor complex where it signals, but with a coreceptor contributing to ligand binding (e.g., in analogy to Smoothened) and its regulation by the Hedgehog binding Patched receptor in Hedgehog signal transduction (e.g., reviewed in Ingham, 1998; Johnson and Scott, 1998).

Nevertheless, evidence for an involvement of Wnts in PCP generation comes from the analysis of convergent extension (CE) in vertebrates. In Xenopus and zebrafish, Wnt11/*silberblick* has been identified as a factor regulating CE (Heisenberg et al., 2000; Tada and Smith, 2000). Similarly, Wnt5 has been implicated in Xenopus (Moon et al., 1993) and zebrafish gastrulation (*pipetail*) (Lele et al., 2001), suggesting that these Wnt family members are specifically required in vertebrate convergent extension. The molecular identification of zebrafish *knypek* as a glypican similar to *Drosophila dally* (Topczewski et al., 2001), further supports the importance of Wnts in CE. The existing data in vertebrates, however, argues for a permissive rather than an instructive role of the Wnts during CE. For example, Wnt11 RNA injections into embryos (at the one- or two-cell stage) rescue the *silberblick* mutant phenotype, indicating that localized expression may not be required for Wnt11 function (Heisenberg et al., 2000). The expression patterns of Wnt11 or Wnt5 also do not easily fit a patterning role during CE (Moon et al., 1993; Heisenberg et al., 2000; Tada and Smith, 2000) (see chapter by Jessen et al. in this book for more details). None of the Wnts in the *Drosophila* genome are orthologues of Wnt11 or Wnt5.

Are there any upstream regulators of *fz*–PCP signaling in *Drosophila*? Although no clear activating factor has been identified to date, recent papers on PCP establishment in the eye suggest that the proto-cadherins *fat* and Dachsous (*ds*) might play a role in regulating *fz* activity (Rawls et al., 2002; Yang et al., 2002). Although *fat* and *ds* are not required for *fz* activation *per se* (*fz*–PCP signaling appears largely normal in these mutants) they regulate the *fz* signaling bias within the R3/R4 pair cell fate decision (Yang et al., 2002). *fat* might act as a positive regulator of *fz*, and the related *ds* as a *fat* antagonist. Moreover, *fat* and *ds* interact genetically with *four jointed* (*fj*) (Yang et al., 2002), which was also suggested to regulate *fz*–PCP activity (Zeidler et al., 1999; Zeidler et al., 2000). It has thus been proposed that *fj* and *ds* act on *fat*, which in turn positively regulates *fz* activity (Yang et al., 2002). This model is deduced from work in the eye, but might be of general importance as *fat, ds*, and *fj* display a PCP requirement in all tissues (Adler et al., 1998; Zeidler et al., 2000; Ma et al., 2003) (see also chapter by Strutt & Strutt in this book).

Other analyses of the roles of *fat* and *ds* in the eye have, however, led to other interpretations of the data. Rawls et al. (Rawls et al., 2002) place *fat* and *ds* further upstream of *fz*–PCP signaling, and they propose a more global role in the establishment of the general polarization axis, preceding *fz*–PCP signaling. Thus, as intriguing as these cadherins are as mediators of a long-range patterning element, the direct role(s) of *fat* and *ds* in *fz*–PCP regulation remains unclear.

A recent functional analysis of *four jointed* (*fj*) has revealed that it might act as an enzyme in the Golgi (Strutt et al., 2004). As *fj* encodes a type II transmembrane protein that is expressed in a D/V gradient in the eye imaginal disc (Zeidler et al., 1999), it has been proposed that it might regulate the activity of one of the core PCP genes, or possibly *ds* (Strutt et al., 2004). The expression of *fj* is regulated by the Notch, JAK/STAT, and Wingless pathways consistent with the idea that it mediates their effects in D/V patterning and retinal polarity (Zeidler et al., 1999). Loss-of-function clones and ectopic expression analyses of *fj* have revealed nonautonomous defects in ommatidial polarity within the D/V axis, as one might expect for the presumed factor X (Wehrli and Tomlinson, 1998). However, the complete removal of *fj* function results in only very mild polarity defects (Zeidler et al., 1999). These observations suggest that *fj* participates redundantly in D/V polarization, possibly by modulating either the activity or stability of *ds* (Strutt et al., 2004), or more directly, the activity of the *fz* receptor itself. Biochemical and genetic experiments to identify the genes and proteins *fj* interacts with will be necessary to resolve these issues.

3.4. Regulation of frizzled activity: intracellular events

Although the *fz–dsh* cascade (Fig. 3) has received a lot of attention, many "primary" PCP genes do not fall directly into this pathway. The link and/or regulatory inputs of these PCP genes (Table 1) to *fz*–PCP signaling are not yet resolved and this is an active area of research. All these proteins display asymmetric protein localization in wing cells in the proximal-distal axis, and within the dorso-ventral axis in the eye at the end of PCP signaling and their interactions (Usui et al., 1999; Axelrod, 2001; Feiguin et al., 2001; Das et al., 2002; Strutt et al., 2002; Tree et al., 2002; Yang et al., 2002; Bastock et al., 2003; Jenny et al., 2003; Rawls and Wolff, 2003; Das et al., 2004). The specific subcellular localization in wing cells (proximal or distal) also correlates with the genetic requirement in the R3/R4 cell pair in the eye (Table 1). These PCP genes include: *pk*, coding for a cytoplasmic protein with several conserved protein–protein interaction motifs (3 LIM domains and a conserved PET (Prickle-Espinas-Testin) domain) (Gubb et al., 1999); *fmi/stan*, a seven-pass transmembrane receptor-like Cadherin (Chae et al., 1999; Usui et al., 1999); *stbm/Vang*: a 4-pass transmembrane protein, with no homology to domains with a defined function (Taylor et al., 1998; Wolff and Rubin, 1998); and *dgo*, a cytoplasmic protein with 6 Ankyrin repeats (Feiguin et al., 2001; Das et al., 2004) (Table 1). All these are localized to either the proximal, distal, or both ends of the cell during PCP establishment (Table 1) and appear to form a complex interaction network with each other and with *fz-dsh* (Fig. 4).

New insights into this network have just emerged. Recent studies have begun to define the role of pk and *stbm/Vang* (Tree et al., 2002; Bastock et al., 2003; Jenny et al., 2003). In brief, pk is recruited to the membrane by *stbm/Vang* by direct molecular interaction, and can there cluster *stbm-pk* complexes via interactions with itself (Bastock et al., 2003; Jenny et al., 2003). Although, *stbm/Vang* mutants show similar phenotypic features as *fz* and *dsh* (Taylor et al., 1998; Wolff and Rubin, 1998), their cellular requirement in the eye is opposite to that of *fz/dsh*: *stbm* is genetically required in R4 and not in R3 (Wolff and Rubin, 1998). Similarly, although both *fz* and *stbm/Vang* mutant clones affect wild-type cells nonautonomously in the wing: *stbm/Vang* affects the proximal side of a clone whereas *fz* affects the distal side (Vinson and Adler, 1987; Taylor et al., 1998). Thus it was proposed that the role of *stbm/Vang* might be to antagonize *fz* signaling. Taken together with the recent molecular data, these observations indicate that the *stbm/Vang–pk* complex antagonizes *fz–dsh* signaling by *pk* binding (and inhibiting) *dsh* (Tree et al., 2002). The

Table 1
Core PCP genes in *Drosophila* eye patterning (all have vertebrate homologues with an equivalent function)

PCP gene	Tissues gene affects in *Drosophila*	Molecular features	Localization in wing cells	R3/R4 req. in eye
Frizzled (fz)	All adult tissues	Seven-pass transmembrane receptor, binds Wg/Wnt ligands, binds Dsh	Distal	R3
Disheveled (dsh)	All adult tissues	Cytoplasmic protein containing DIX, PDZ, DEP domains, recruited to membrane by Fz, binds Fz, Pk, Stbm and Dgo	Distal	R3
Prickle (pk) (a.k.a. *prickle-spiny legs*)	All adult tissues	Cytoplasmic protein with 3 LIM domains and PET domain, recruited to membrane by Stbm, physically interacts with Dsh, Stbm and Dgo	Proximal	R4
Strabismus (stbm)/ *Van Gogh (Vang)*	All adult tissues	Novel 4-pass transmembrane protein binds Pk, Dsh and Dgo	Proximal	R4
Flamingo (fmi)/starry night (stan)	All adult tissues	Cadherin with seven-pass transmembrane receptor features	Proximal + distal	R3 + R4
Diego (dgo)	Eye, wing, notum in GOF[a]	Cytoplasmic Ankyrin repeat protein, recruited to membrane by Fz, binds Dsh, Stbm and Dgo	Distal; co-loc. with Fz/Dsh	R3?
RhoA	Eye, wing[a]	Small GTPase	n.d.	R3
Misshapen (msn)	Eye, wing, notum[a]	STE20-like S/T protein kinase	n.d.	R3
Fat (Ft)	All adult tissues	Proto-cadherin, interacts with Ds, binds Atrophin	Distal	R3
Dachsous (ds)	All adult tissues	Proto-cadherin, interacts with Fat	Proximal	R4
Four jointed (fj)	All adult tissues	Type-2 transmembrane or secreted peptide possibly functions in Golgi to modify Ds	?	

[a]Other tissues were not tested.

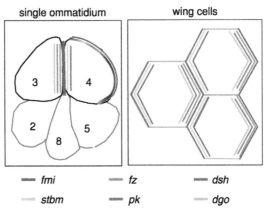

Fig. 4. Subcellular localization of PCP factors in the eye and wing. In the *Drosophila* eye, the subcellular localization differences of PCP proteins are restricted to the R3/R4 precursor cells (left panel), whereas in the wing they are apparent in all cells of the differentiating wing field (right panel for comparison). Initially, all PCP proteins are localized around the whole apical cortex (not shown) and following their interactions become restricted to either the R3 side or the R4 side of the R3/R4 cell boundary. The color code for the PCP factors is as shown. Note that *fmi* (purple) localizes to both sides of the R3/R4 boundary within the DV axis (and both sides of each wing cell boundary in the proximo-distal axis). The functional significance for the enrichment of all PCP factors around the R4 precursor cortex (shown in rainbow colors) is not known, and there is no genetic evidence for a requirement of *fz, dsh,* or *dgo* in R4. The localization of *dsh* and *pk* is deduced from their genetic requirements, parallel localizations in wing cells, and stable physical interactions with either *fz* (for *dsh*) or *stbm* (*pk*). (See Color Insert.)

precise mechanism of the *dsh* inhibition is not yet clear, but it causes an absence of *dsh* at the membrane, possibly mediated through local endocytosis or degradation of *dsh*. The general features of these interactions are largely conserved between the eye, wing, abdomen, and notum of *Drosophila* (Rawls et al., 2002; Tree et al., 2002; Bastock et al., 2003; Jenny et al., 2003; Bellaiche et al., 2004; Lawrence et al., 2004), and thus this is likely to be a general mechanism used in all PCP contexts and also in vertebrates (e.g., Darken et al., 2002; Jessen et al., 2002; Veeman et al., 2003).

This model is indeed supported by data from recent work in vertebrates (Darken et al., 2002; Goto and Keller, 2002; Jessen et al., 2002; Park and Moon, 2002). In particular, Xenopus *stbm* (X*stbm*) has an antagonistic behavior to Wnt–*fz* signaling and a physical interaction between *stbm* and X*dsh* has been reported (Park and Moon, 2002). These data are consistent between flies and vertebrates. However, the observation that X*stbm* can also activate JNK signaling in cell culture (Park and Moon, 2002), the opposite of what would be expected from the genetic analyses in flies, suggests that the interplay between these PCP genes might be more complex and context dependent. Nevertheless, the recent addition of the PCP genes *strabismus* and *prickle* to the convergent extension process (Bekman and Henrique, 2002; Darken et al., 2002; Goto and Keller, 2002; Jessen et al., 2002; Park and Moon, 2002; Wallingford et al., 2002; Carreira-Barbosa et al., 2003; Katoh, 2003; Takeuchi et al., 2003) strongly suggests that most (if not all) of the PCP genes interact, and that the resulting signaling cassette is evolutionarily conserved.

However, many mechanistic questions remain. How does *pk* prevent *dsh* membrane localization when it is itself localized to the membrane? Does it render *dsh* unstable? It is noteworthy that, although the signaling differences are most clearly manifest in the distinct subcellular localization of these proteins, the actual signaling events must initiate earlier when all these factors are co-localized around the whole apical cortex of the respective cells (see references above).

The role of the Cadherin *fmi–stan* (Chae et al., 1999; Usui et al., 1999; Das et al., 2002) is intriguing but less clear. *fmi* co-localizes with both *fz–dsh* distally and with *stbm–pk* proximally in wing cells, and at both sides of the R3/R4 cell boundary in the eye (Chae et al., 1999; Usui et al., 1999; Das et al., 2002). It appears to be required for *fz–dsh* localization (Usui et al., 1999; Das et al., 2002) as well as for *stbm–pk* localization (Tree et al., 2002) (see also Fig. 4). The functional analysis of *fmi* during eye PCP establishment (Das et al., 2002) also supports a dual role for *fmi* in PCP establishment. *fmi* is required in both cells of the R3/R4 pair, and although *fmi* displays homophilic cell-adhesion behavior (Usui et al., 1999), offering a simple explanation for the requirement in both cells, it also has distinct functions in each cell of the R3/R4 pair (Das et al., 2002). Strikingly, *fmi* appears to inhibit *fz–dsh* signaling in R4 (Das et al., 2002) (possibly through stabilizing the *stbm–pk* complex; see also Fig. 4). This model, with an inhibitory role of *fmi* on *fz* activity, was recently also supported by the analysis of PCP establishment on the abdominal cuticle (Lawrence et al., 2004). Furthermore, overexpression of *fmi* in the wing also has an antagonistic effect on *fz* signaling (Taylor et al., 1998; Usui et al., 1999). Thus *fmi* might have dual functions in the R3 and R4 cells in the eye, and on the distal and proximal sides of wing or abdominal cells, possibly mediated by distinct *fmi*-containing protein complexes (see Fig. 4 for model).

What could be the mechanistic basis of the distinct *fmi* functions on different sides of a cell? Very little insight exists in this question as so far no physical interactions between *fmi* and any other protein have been reported. Nevertheless, as a Cadherin family member *fmi* could interact extracellularly with other cadherins, notably the PCP factors *fat* and/or *ds*, leading to changes in its activity through these adhesive interactions. Alternatively, intracellular interactions could influence its activity. This could possibly be achieved through *dgo*, which co-localizes with *fmi* (Feiguin et al., 2001) and interacts with it genetically (Das et al., 2002). Overexpression of *dgo* in the wing shows a phenotype that is indistinguishable from overexpressed *fz*, suggesting that *dgo* acts positively on *fz* signaling (Feiguin et al., 2001). Thus, the presence or absence of *dgo* might affect the function of *fmi*.

4. How is PCP signaling interpreted within each ommatidium?

How does *fz*–PCP signaling regulate the polarization of a group of cells, like the ommatidial cluster in the eye? Genetic manipulation has shown that relative *fz* activity in the R3/R4 photoreceptor pair is critical for this process. In particular, the cell that has higher *fz* activity (or might acquire it first in a wild-type background) will always become the R3 cell, and this in turn will induce the neighboring cell as the R4 photoreceptor, giving an ommatidium the respective chirality (Zheng et al., 1995; Fanto and Mlodzik, 1999; Tomlinson and Struhl, 1999). However, R3 and R4 precursors are direct neighbors within the 5-cell precluster and can be over a hundred cells away from the presumptive source of the signal. Consequently, the difference in *fz* activity in these two cells should be very small. It is thus a difficult task to establish a 100% reliable read-out between the R3/R4 cells. In addition, in *fz* and *dsh* mutants, most ommatidia still remain chiral with R3/R4 being specified in these clusters, albeit stochastically.

These observations suggested that a secondary signal, acting downstream of *fz*, is required for R3/R4 cell fate specification. The Notch pathway has been implicated as this secondary signal in the generation of chirality (Fig. 5). In particular, Notch signaling specifies the R4 photoreceptor (Cooper and Bray, 1999; Fanto and Mlodzik, 1999; Tomlinson and Struhl, 1999) (summarized in Fig. 4). Expression analysis of the membrane-associated Notch ligand, *Delta*, demonstrates that *Delta* is a transcriptional target of *fz* signaling in the R3 precursor (Cooper and Bray, 1999; Fanto and Mlodzik, 1999). *fz* and several downstream components of *fz*, including *dsh*, Rac, *rhoA*, Hep/JNKK, and the transcription factor dJun regulate *Delta* expression within the R3/R4 pair (Fanto and Mlodzik, 1999; Fanto et al., 2000; Weber et al., 2000).

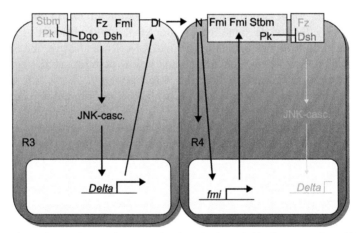

Fig. 5. *fz*–PCP factor interactions and the two-tiered mechanism of chirality generation within the R3/R4 pair. An initial difference in *fz–dsh*–PCP signaling levels, as generated through the interactions between *fz–dsh* and the other PCP factors, is amplified by the transcriptional upregulation of *Delta* and the subsequent activation of Notch signaling in R4. Notch activation in turn inhibits Delta transcription in R4, further increasing the *fz*–PCP signaling difference between the R3 and R4 cells and thus creating a very solid binary cell fate decision. In mutants of PCP factors the initial asymmetry of *Delta* upregulation is lost and a stochastic Delta/Notch interaction generates randomly the R3/R4 cell fates and chirality. Within the PCP gene cassette the role of *stbm/Vang* and *prickle* (*pk*) is to inhibit *fz–dsh* signaling. Thus *stbm* and *pk* are required mainly in the R4 cell where they antagonize the *fz*–PCP pathway. *fmi* appears to be required to stabilize both membrane-associated complexes, although the mechanisms of its action remains unknown as no physical interactions between *fmi* and any other PCP gene have been reported. The role of Diego is also not yet clear, except that it is positively required for *fz*–PCP signaling. See text for details. (See Color Insert.)

Following its transcriptional-*fz*-signaling-mediated upregulation, *Delta* amplifies the differences in *fz* signaling by activating Notch in the R4 precursor, thus locking the binary R3/R4 cell fate and chirality decision in place (Cooper and Bray, 1999; Fanto and Mlodzik, 1999; Tomlinson and Struhl, 1999). The signal amplification is even enhanced by the fact that activated Notch signaling represses *Delta* expression in R4 (summarized in Fig. 5). This two-tiered mechanism explains how a small initial difference in *fz* signaling between the R3 and R4 precursor cells reliably generates the correct decision. This observation also explains why chirality, albeit stochastically, is still generated in *fz* and *dsh* null mutants.

The input from cell type-specific transcription factors to the R3/R4 cell fate determination remains a largely open question. As mentioned above, two sets of transcription factors have been implicated in R3 and R4 fate induction and differentiation. These are Seven-up (Mlodzik et al., 1990) and Spalt (Domingos et al., 2004). Both of these are required for the R3/R4 cell fates in general, and in the respective mutants the R3/R4 precursors become transformed into R7-type photoreceptors (Mlodzik et al., 1990; Domingos et al., 2004). However, the activity of both these factors is also a prerequisite for the proper interpretation (and/or activity regulation) of the *fz*–PCP pathway (Domingos et al., 2004; Fanto et al., 1998) and in the case of Spalt, also the resulting Notch response in R4 (Domingos et al., 2004). It is likely that other cell type-specific transcription factors are induced in response to either the *fz*–PCP pathway or Notch signaling in the R3/R4 precursors.

5. Ommatidial rotation

Besides the establishment of ommatidial chirality through the R3/R4 cell fate specification, the subsequent rotation of each ommatidium is a key aspect of PCP establishment in the *Drosophila* eye (Fig. 1). Despite the recent insights into the *fz*–PCP signaling pathway and its role in R3/R4 cell fate induction, very little is still known about the associated ommatidial rotation. The rotation is an important PCP aspect in the retina, and a fascinating biological problem, as groups of cells (the entire ommatidial precluster) have to rotate as a whole within the epithelium. Strikingly, rotation initiates at a point when some of the photoreceptor precursors (R8, R2/R5) have already initiated axonal outgrowth towards the optic lobes in the brain (Wolff and Ready, 1993).

In all mutants of the primary PCP genes, rotation is still taking place, even though often at random, indicating that the two processes, determination of polarity/chirality and subsequent rotation, are (at least in part) independent. There are few genes known that affect *only* the rotation process. These are mainly *roulette* (*rlt*) and *nemo* (Choi and Benzer, 1994), and *dRok*, the *rho*-associated kinase, acting as an effector of *rhoA* in PCP signaling (Winter et al., 2001). Interestingly, *rlt* and *nemo* have opposite effects on the rotation process. Whereas, in *nemo* mutants ommatidia generally under-rotate, stopping after the first 45°, in the *rlt* mutant ommatidia fail to stop at 90° (Choi and Benzer, 1994), suggesting that *nemo* is positively required in the process, whereas *rlt* acts negatively. The *nemo, roulette* double mutant displays largely the same defect as *nemo* alone, suggesting that *nemo* acts upstream of *rlt* in this process. Mutant alleles of *dRok* show variable rotation defects (Winter et al., 2001).

Of these three genes only *dRok* is linked to *fz*–PCP signaling and is likely to act there as an effector of *rhoA* (Winter et al., 2001). *nemo* is a distant member of the MAPK super-family (Choi and Benzer, 1994). There is no clear link between the *fz*–PCP pathway and *nemo*, although it has been shown that *fz–nemo* double mutants produce a more severe phenotype than either single mutant. Specifically, in a *fz–nemo* double mutant, many ommatidia do not rotate at all (Zheng et al., 1995), suggesting that *fz* has some positive effect on *nemo* activity. In other contexts, the *nemo*-like kinases have been shown to act as antagonists of canonical Wnt/β-catenin signaling (e.g., Ishitani et al., 1999; Meneghini et al., 1999).

The third gene, *rlt*, has been the least studied and least understood until recently. However, three studies published in 2003 put *rlt* on the map (Brown and Freeman, 2003; Gaengel and Mlodzik, 2003; Strutt and Strutt, 2003). Interestingly, the *rlt* mutation is an allele of the inhibitory ligand *Argos* of the EGF-receptor (Egfr). Thus, this discovery implicates Egfr-Ras signaling in the regulation of ommatidial rotation. Indeed several hypomorphic alleles of Egfr itself and other pathway components display rotation defects (Brown and Freeman, 2003; Gaengel and Mlodzik, 2003). As Egfr–Ras signaling plays a crucial role in many aspects of eye development ranging from proliferation to survival of specified cells (e.g., Zipursky and Rubin, 1994; Bergmann et al., 1998; Freeman, 1998; Casci and Freeman, 1999; Van Buskirk and Schupbach, 1999), it is difficult to dissect the role of the pathway in rotation. Nevertheless, the existing data suggest that downstream of Ras, there is a bifurcation of Egfr signaling and both the MAPK and the Canoe/AF-6 effectors are required (Gaengel and Mlodzik, 2003). Genetic interaction studies and expression analyses suggest that that Egfr–Ras signaling might affect cadherin function and or localization in this context (Brown and Freeman, 2003; Gaengel and Mlodzik, 2003). Although no clear link between the Egfr and *fz*–PCP pathways are known, the activity of Egfr-signaling appears to affect the localization of *fz* itself (Strutt and Strutt, 2003).

As the rotation process *per se* implies that ommatidial precursor cells have to change their adhesive properties in the process, it is an interesting question whether and what cell adhesion features and molecules are involved in the context. Interestingly, besides the potential involvement of cadherins (as suggested by their interactions with Egfr pathway, see above) (Brown and Freeman, 2003; Gaengel and Mlodzik, 2003), another cell adhesion factor with a defect in rotation is *LamininA* (Henchcliffe et al., 1993). In *LamininA* mutants the rotation appears random, whereas the chirality is not affected. However, the function of Laminin has not been linked to any other gene/molecule in this context and thus this remains just an interesting observation at this point.

6. General conclusions

What are the major lessons and evolutionary generalities that we can learn from the establishment of tissue polarity in the retina? The discovery of the *fz*–PCP pathway, as an alternative effector pathway of Wnt-mediated *fz–dsh* activation (see

Fig. 3), in the generation of PCP is a very important new insight. As is now obvious, this *fz*–PCP signaling pathway is conserved in evolution in most (if not all) animals and PCP contexts, and for example is also used in the coordination of cellular migrations in vertebrate convergent extension during gastrulation (reviewed in e.g., Keller, 2002; Myers et al., 2002; Tada et al., 2002, see also chapter by Jessen et al. in this book). Similarly, Wnt–*fz* signaling has been implicated in many contents throughout the animal kingdom, ranging from embryonic patterning to proliferation control. Thus, studies of retinal polarity and patterning are relevant models for many other developmental processes and human disease states.

The two-tiered *fz*–Notch activation to polarize groups of cells might be a general mechanism for PCP establishment in multicellular units. The asymmetric activation of Notch-signaling in response to a graded signal is possibly used also in the generation of polarity in the multicellular feather bud precursors in vertebrates, where polarized expression of Notch pathway components has been reported (Chen et al., 1997). This appears reminiscent of the regulation of *Delta* expression by the *fz*–PCP signaling pathway in the *Drosophila* eye. Although it is not yet known whether *fz* signaling is involved in the context of feather bud polarization, the usage of *fz* and Notch in the same processes is observed in several other tissues (e.g., bristle patterning and wing development in *Drosophila*). The potentially antagonistic or synergistic *fz*–Notch interactions have led to speculations that Notch is also involved in Wg signal transmission (Couso and Martinez Arias, 1994). The observation that *dsh* can interact directly with Notch as shown in a yeast two-hybrid assay (Axelrod et al., 1996) suggests that the pathway interactions could be complex, and a lot more work will be required to be able to understand the specific molecular and cell biological features of PCP signaling in the fly eye and in general.

Acknowledgments

I am grateful to all members of the Mlodzik lab and many colleagues in the field of PCP signaling and Wnt signaling in general for their helpful and stimulating discussions. Eye specific work in the Mlodzik lab is supported by NIH-NEI grants EY13256 and EY 14597.

References

Adler, P.N. 1992. The genetic control of tissue polarity in *Drosophila*. BioEssays 14, 735–741.

Adler, P.N. 2002. Planar signaling and morphogenesis in *Drosophila*. Dev. Cell 2, 525–535.

Adler, P.N., Charlton, J., Liu, J. 1998. Mutations in the cadherin superfamily member gene dachsous cause a tissue polarity phenotype by altering *frizzled* signaling. Development 125, 959–968.

Adler, P.N., Vinson, C., Park, W.J., Conover, S., Klein, L. 1990. Molecular structure of *frizzled,* a *Drosophila* tissue polarity gene. Genetics 126, 401–416.

Axelrod, J., Matsuno, K., Artavanis-Tsakonas, S., Perrimon, N. 1996. Interaction between Wingless and Notch signaling pathways mediated by Disheveled. Science 271, 1826–1832.

Axelrod, J.D. 2001. Unipolar membrane association of Disheveled mediates *frizzled* planar cell polarity signaling. Genes Dev. 15, 1182–1187.

Axelrod, J.D., Miller, J.R., Shulman, J.M., Moon, R.T., Perrimon, N. 1998. Differential requirement of Disheveled provides signaling specificity in the Wingless and planar cell polarity signaling pathways. Genes Dev. 12, 2610–2622.

Bastock, R., Strutt, H., Strutt, D. 2003. *Strabismus* is asymmetrically localised and binds to *Prickle* and *Disheveled* during *Drosophila* planar polarity patterning. Development 130, 3007–3014.

Bekman, E., Henrique, D. 2002. Embryonic expression of three mouse genes with homology to the *Drosophila* melanogaster prickle gene. Mech. Dev. 119(Suppl. 1), S77–S81.

Bellaiche, Y., Beaudoin-Massiani, O., Stuttem, I., Schweisguth, F. 2004. The planar cell polarity protein Strabismus promotes Pins anterior localization during asymmetric division of sensory organ precursor cells in *Drosophila*. Development 131, 469–478.

Bergmann, A., Agapite, J., McCall, K., Steller, H. 1998. The *Drosophila* gene hid is a direct molecular target of Ras-dependent survival signaling. Cell 95, 331–341.

Bhanot, P., Brink, M., Samos, C.H., Hsieh, J.-C., Wang, Y., Macke, J.P., Andrew, D., Nathans, J., Nusse, R. 1996. A new member of the *frizzled* family from *Drosophila* functions as a Wingless receptor. Nature 382, 225–230.

Bhanot, P., Fish, M., Jemison, J.A., Nusse, R., Nathans, J., Cadigan, K.M. 1999. *frizzled* and D *frizzled*-2 function as redundant receptors for wingless during *Drosophila* embryonic development. Development 126, 4175–4186.

Bhat, K.M. 1998. *frizzled* and *frizzled* 2 play a partially redundant role in Wingless signaling and have similar requirements to Wingless in neurogenesis. Cell 95, 1027–1036.

Blair, S. 1999. Eye development: Notch lends a handedness. Curr. Biol. 9, 356–360.

Boutros, M., Mlodzik, M. 1999. Disheveled: At the crossroads of divergent intracellular signaling pathways. Mech. Dev. 83, 27–37.

Boutros, M., Paricio, N., Strutt, D.I., Mlodzik, M. 1998. Disheveled activates JNK and discriminates between JNK pathways in planar polarity and *wingless* signaling. Cell 94, 109–118.

Brown, K.E., Freeman, M. 2003. Egfr *signaling* defines a protective function for ommatidial orientation in the *Drosophila* eye. Development 130, 5401–5412.

Carreira-Barbosa, F., Concha, M.L., Takeuchi, M., Ueno, N., Wilson, S.W., Tada, M. 2003. Prickle 1 regulates cell movements during gastrulation and neuronal migration in zebrafish. Development 130, 4037–4046.

Casci, T., Freeman, M. 1999. Control of EGF receptor *signaling*: Lessons from fruitflies. Cancer Metastasis Rev. 18, 181–201.

Chae, J., Kim, M.J., Goo, J.H., Collier, S., Gubb, D., Charlton, J., Adler, P.N., Park, W.J. 1999. The *Drosophila* tissue polarity gene *starry night* encodes a member of the protocadherin family. Development 126, 5421–5429.

Chen, C.-N., Struhl, G. 1999. Wingless transduction by the *frizzled* and *frizzled*2 proteins of *Drosophila*. Development 126, 5441–5452.

Chen, C.W.J., Jung, H.S., Jiang, T.X., Chuong, C.M. 1997. Asymmetric expression of Notch/Delta/Serrate is associated with the anterior–posterior axis of feather buds. Dev. Biol. 188, 181–188.

Choi, K.-W., Benzer, S. 1994. Rotation of photoreceptor clusters in the developing *Drosophila* eye requires the *nemo* gene. Cell 78, 125–136.

Clandinin, T.R., Zipursky, S.L. 2000. Afferent growth cone interactions control synaptic specificity in the *Drosophila* visual system. Neuron 28, 427–436.

Clandinin, T.R., Zipursky, S.L. 2002. Making connections in the fly visual system. Neuron 35, 827–841.

Cooper, M.T.D., Bray, S.J. 1999. *frizzled* regulation of Notch *signaling* polarizes cell fate in the *Drosophila* eye. Nature 397, 526–529.

Couso, J.P., Martinez Arias, A. 1994. Notch is required for Wingless signaling in *Drosophila*. Cell 79, 259–272.

Curtin, J.A., Quint, E., Tsipouri, V., Arkell, R.M., Cattanach, B., Copp, A.J., Henderson, D.J., Spurr, N., Stanier, P., Fisher, E.M., et al. 2003. Mutation of Celsr1 disrupts planar polarity of inner ear hair cells and causes severe neural tube defects in the mouse. Curr. Biol. 13, 1129–1133.

Dabdoub, A., Donohue, M.J., Brennan, A., Wolf, V., Montcouquiol, M., Sassoon, D.A., Hseih, J.C., Rubin, J.S., Salinas, P.C., Kelley, M.W. 2003. Wnt signaling mediates reorientation of outer hair cell stereociliary bundles in the mammalian cochlea. Development 130, 2375–2384.

Darken, R.S., Scola, A.M., Rakeman, A.S., Das, G., Mlodzik, M., Wilson, P.A. 2002. The planar polarity gene *strabismus* regulates convergent extension movements in Xenopus. EMBO J. 21, 4409–4420.

Das, G., Jenny, A., Klein, T.J., Eaton, S., Mlodzik, M. 2004. Diego interacts with Prickle and Strabismus/Van Gogh to localize planar cell polarity complexes. Development 131, 4467–4476.

Das, G., Reynolds-Kenneally, J., Mlodzik, M. 2002. The Atypical Cadherin Flamingo Links *frizzled* and Notch Signaling in Planar Polarity Establishment in the *Drosophila* Eye. Dev Cell 2, 655–666.

Dietrich, W. 1909. Die Facettenaugen der Dipteren. Z Wiss. Zool. 92, 465–539.

Domingos, P.M., Mlodzik, M., Mendes, C.S., Brown, S., Steller, H., Mollereau, B. 2004. Spalt transcription factors are required for R3/R4 specification and establishment of planar cell polarity in the *Drosophila* eye. Development 131, 5695–5702.

Eaton, S. 1997. Planar polarity in *Drosophila* and vertebrate epithelia. Curr. Op. Cell Biol. 9, 860–866.

Eaton, S., Wepf, R., Simons, K. 1996. Roles for Rac1 and Cdc42 in planar polarization and hair outgrowth in the wing of *Drosophila*. J. Cell Biol. 135, 1277–1289.

Fanto, M., Mayes, C.A., Mlodzik, M. 1998. Linking cell-fate specification to planar polarity: Determination of the R3/R4 photoreceptors is a prerequisite for the interpretation of the *frizzled* mediated polarity signal. Mech. Dev. 74, 51–58.

Fanto, M., Mlodzik, M. 1999. Asymmetric Notch activation specifies photoreceptors R3 and R4 and planar polarity in the *Drosophila* eye. Nature 397, 523–526.

Fanto, M., Weber, U., Strutt, D.I., Mlodzik, M. 2000. Nuclear signaling by Rac abd Rho GTPases is required in the establishment of epithelial planar polarity in the *Drosophila* eye. Curr. Biol. 10, 679–688.

Feiguin, F., Hannus, M., Mlodzik, M., Eaton, S. 2001. The Ankyrin repeat protein Diego mediates *frizzled*-dependent planar polarization. Developmental Cell 1, 93–101.

Freeman, M. 1998. Complexity of EGF receptor *signaling* revealed in *Drosophila*. Curr. Opin. Genet. Dev. 8, 407–411.

Gaengel, K., Mlodzik, M. 2003. Egfr signaling regulates ommatidial rotation and cell motility in the *Drosophila* eye via MAPK/Pnt signaling and the Ras effector Canoe/AF6. Development 130, 5413–5423.

Gho, M., Schweisguth, F. 1998. *frizzled* signaling controls orientation of asymetric sense organ precursor cell divisions in *Drosophila*. Nature 393, 178–181.

Goto, T., Keller, R. 2002. The planar cell polarity gene strabismus regulates convergence and extension and neural fold closure in Xenopus. Dev. Biol. 247, 165–181.

Gubb, D. 1993. Genes controlling cellular polarity in *Drosophila*. Development (Suppl. 1993), 269–277.

Gubb, D., Green, C., Huen, D., Coulson, D., Johnson, G., Tree, D., Collier, S., Roote, J. 1999. The balance between isoforms of the prickle LIM domain protein is critical for planar polarity in *Drosophila* imaginal discs. Genes Dev. 13, 2315–2327.

Hakeda-Suzuki, S., Ng, J., Tzu, J., Dietzl, G., Sun, Y., Harms, M., Nardine, T., Luo, L., Dickson, B.J. 2002. Rac function and regulation during *Drosophila* development. Nature 416, 438–442.

Heberlein, U., Moses, K. 1995. Mechanisms of *Drosophila* retinal morphogenesis: The virtues of being progressive. Cell 81, 987–990.

Heisenberg, C.P., Tada, M., Rauch, G.J., Saude, L., Concha, M.L., Geisler, R., Stemple, D.L., Smith, J.C., Wilson, S.W. 2000. Silberblick/Wnt11 mediates convergent extension movements during zebrafish gastrulation. Nature 405, 76–81.

Henchcliffe, C., Garcia-Alonso, L., Tang, J., CS, G. 1993. Genetic analysis of laminin A reveals diverse functions during morphogenesis in *Drosophila*. Development 118, 325–337.

Ingham, P.W. 1998. Transducing Hedgehog: The story so far. EMBO J. 17, 3505–3511.

Ishitani, T., Ninomiya-Tsuji, J., Nagai, S., Nishita, M., Meneghini, M., Barker, N., Waterman, M., Bowerman, B., Clevers, H., Shibuya, H., Matsumoto, K. 1999. The TAK1-NLK-MAPK-related pathway antagonizes *signaling* between beta-catenin and transcription factor TCF. Nature 399, 798–802.

Jenny, A., Darken, R.S., Wilson, P.A., Mlodzik, M. 2003. Prickle and Strabismus form a functional complex to generate a correct axis during planar cell polarity signaling. EMBO J. 22, 4409–4420.

Jessen, J.R., Topczewski, J., Bingham, S., Sepich, D.S., Marlow, F., Chandrasekhar, A., Solnica-Krezel, L. 2002. Zebrafish trilobite identifies new roles for Strabismus in gastrulation and neuronal movements. Nat. Cell Biol. 4, 610–615.

Johnson, R.L., Scott, M.P. 1998. New players and puzzles in the Hedgehog signaling pathway. Curr. Opin. Genet. Dev. 8, 450–456.

Jones, K. H.J.L., Adler, P.N. 1996. Molecular analysis of EMS-induced frizzled mutations in *Drosophila melanogaster*. Genetics 142, 205–215.

Katoh, M. 2003. Identification and characterization of human PRICKLE1 and PRICKLE2 genes as well as mouse Prickle1 and Prickle2 genes homologous to *Drosophila* tissue polarity gene prickle. Int. J. Mol. Med. 11, 249–256.

Keller, R. 2002. Shaping the vertebrate body plan by polarized embryonic cell movements. Science 298, 1950–1954.

Kennerdell, J.R., Carthew, R.W. 1998. Use of dsRNA-mediated genetic interference to demonstrate that *frizzled* and *frizzled2* act in the *wingless* pathway. Cell 95, 1017–1026.

Klingensmith, J., Nusse, R., Perrimon, N. 1994. The *Drosophila* segment polarity gene *disheveled* encodes a novel protein required for response to the *wingless* signal. Genes Dev. 8, 118–130.

Krasnow, R.E., Adler, P.N. 1994. A single *frizzled* protein has a dual role in tissue polarity. Development 120, 1883–1893.

Krasnow, R.E., Wong, L.L., Adler, P.N. 1995. *disheveled* is a component of the *frizzled signaling* pathway in *Drosophila*. Development 121, 4095–4102.

Lawrence, P.A., Casal, J., Struhl, G. 2004. Cell interactions and planar polarity in the abdominal epidermis of *Drosophila*. Development 131, 4651–4664.

Lawrence, P.A., Shelton, P.M.J. 1975. The determination of polarity in the developing insect retina. J. Embryol. Exp. Morph. 33, 471–486.

Lele, Z., Bakkers, J., Hammerschmidt, M. 2001. Morpholino phenocopies of the swirl, snailhouse, somitabun, minifin, silberblick, and pipetail mutations. Genesis 30, 190–194.

Li, L., Yuan, H., Xie, W., Mao, J., Caruso, A.M., McMahon, A., Sussman, D.J., Wu, D. 1999. Disheveled proteins lead to two *signaling* pathways: regulation of Lef-1 and c-Jun N-terminal kinase in mammalian cells. J. Biol. Chem. 274, 129–134.

Ma, D., Yang, C.H., McNeill, H., Simon, M.A., Axelrod, J.D. 2003. Fidelity in planar cell polarity *signaling*. Nature 421, 543–547.

Meneghini, M.D., Ishitani, T., Carter, J.C., Hisamoto, N., Ninomiya-Tsuji, J., Thorpe, C.J., Hamill, D.R., Matsumoto, K., Bowerman, B. 1999. MAP kinase and Wnt pathways converge to downregulate an HMG-domain repressor in Caenorhabditis elegans. Nature 399, 793–797.

Mlodzik, M. 1999. Planar polarity in the *Drosophila* eye: A multifaceted view of signaling specificity and cross-talk. EMBO J. 24, 6873–6879.

Mlodzik, M. 2000. Spiny legs and prickled bodies: new insights and complexities in planar polarity establishment. Bioessays 22, 311–315.

Mlodzik, M. 2002. Planar cell polarization: Do the same mechanisms regulate *Drosophila* tissue polarity and vertebrate gastrulation? Trends Genet. 18, 564–571.

Mlodzik, M., Hiromi, Y., Weber, U., Goodman, C.S., Rubin, G.M. 1990. The *Drosophila seven-up* gene, A member of the steroid receptor gene superfamily, controls photoreceptor cell fates. Cell 60, 211–224.

Montcouquiol, M., Rachel, R.A., Lanford, P.J., Copeland, N.G., Jenkins, N.A., Kelley, M.W. 2003. Identification of Vangl2 and Scrb1 as planar polarity genes in mammals. Nature 423, 173–177.

Moon, R.T., Campbell, R.M., Christian, J.L., McGrew, L.L., Shih, J., Fraser, S. 1993. Xwnt-5A: a maternal Wnt that affects morphogenetic movements after overexpression in embryos of Xenopus laevis. Development 119, 97–111.

Mueller, H., Samanta, R., Wieschaus, E. 1999. Wingless signaling in the *Drosophila* embryo: Zygotic requirements and the role of the *frizzled* genes. Development 126, 577–586.

Myers, D.C., Sepich, D.S., Solnica-Krezel, L. 2002. Convergence and extension in vertebrate gastrulae: Cell movements according to or in search of identity? Trends Genetics 18, 447–455.

Paricio, N., Feiguin, F., Boutros, M., Eaton, S., Mlodzik, M. 1999. The *Drosophila* STE20-like kinase Misshapen is required downstream of the *frizzled* receptor in planar polarity signaling. EMBO J. 18, 4669–4678.

Park, M., Moon, R.T. 2002. The planar cell polarity gene *stbm* regulates cell behaviour and cell fate in vertebrate embryos. Nat. Cell Biol. 4, 20–25.

Rawls, A.S., Guinto, J.B., Wolff, T. 2002. The cadherins fat and dachsous regulate dorsal/ventral signaling in the *Drosophila* eye. Curr. Biol. 12, 1021–1026.

Rawls, A.S., Wolff, T. 2003. Strabismus requires Flamingo and Prickle function to regulate tissue polarity in the *Drosophila* eye. Development 130, 1877–1887.

Reifegerste, R., Moses, K. 1999. The genetics of epithelial polarity and pattern in the *Drosophila* retina. BioEssays 21, 275–285.

Strutt, D. 2003. *frizzled signaling* and cell polarisation in *Drosophila* and vertebrates. Development 130, 4501–4513.

Strutt, D., Johnson, R., Cooper, K., Bray, S. 2002. Asymmetric localization of frizzled and the determination of Notch-dependent cell fate in the *Drosophila* eye. Curr. Biol. 12, 813–824.

Strutt, D.I., Weber, U., Mlodzik, M. 1997. The role of *rhoA* in tissue polarity and *frizzled signaling*. Nature 387, 292–295.

Strutt, H., Mundy, J., Hofstra, K., Strutt, D. 2004. Cleavage and secretion is not required for Four-jointed function in *Drosophila* patterning. Development 131, 881–890.

Strutt, H., Strutt, D. 2003. EGF signaling and ommatidial rotation in the *Drosophila* eye. Curr. Biol. 13, 1451–1457.

Strutt, H., Strutt, D.I. 1999. Polarity determination in the *Drosophila* eye. Curr. Opin. Genet. Devel. 9, 442–446.

Tada, M., Concha, M.L., Heisenberg, C.P. 2002. Non-canonical Wnt *signaling* and regulation of gastrulation movements. Semin. Cell Dev. Biol. 13, 251–260.

Tada, M., Smith, J.C. 2000. Xwnt11 is a target of Xenopus Brachyury: Regulation of gastrulation movements via Disheveled, but not through the canonical Wnt pathway. Development 127, 2227–2238.

Takeuchi, M., Nakabayashi, J., Sakaguchi, T., Yamamoto, T.S., Takahashi, H., Takeda, H., Ueno, N. 2003. The prickle-related gene in vertebrates is essential for gastrulation cell movements. Curr. Biol. 13, 674–679.

Taylor, J., Abramova, N., Charlton, J., Adler, P.N. 1998. Van Gogh: A new *Drosophila* tissue polarity gene. Genetics 150, 199–210.

Theisen, H., Purcell, J., Bennett, M., Kansagara, D., Syed, A., Marsh, J.L. 1994. *disheveled* is required during *wingless signaling* to establish both cell polarity and cell identity. Development 120, 347–360.

Tomlinson, A. 1988. Cellular interactions in the developing *Drosophila* eye. Development 104, 183–193.

Tomlinson, A., Ready, D.F. 1987. Cell fate in the *Drosophila* Ommatidium. Dev. Biol. 123, 264–275.

Tomlinson, A., Strapps, W.R., Heemskerk, J. 1997. Linking *frizzled* and Wnt signaling in *Drosophila* development. Development 124, 4515–4521.

Tomlinson, A., Struhl, G. 1999. Decoding vectorial information from a gradient: Sequential roles of the receptors *frizzled* and Notch in establishing planar polarity in the *Drosophila* eye. Development 126, 5725–5738.

Topczewski, J., Sepich, D.S., Myers, D.C., Walker, C., Amores, A., Lele, Z., Hammerschmidt, M., Postlethwait, J., Solnica-Krezel, L. 2001. The zebrafish glypican knypek controls cell polarity during gastrulation movements of convergent extension. Dev. Cell 1, 251–264.

Tree, D.R., Shulman, J.M., Rousset, R., Scott, M.P., Gubb, D., Axelrod, J.D. 2002. Prickle mediates feedback amplification to generate asymmetric planar cell polarity signaling. Cell 109, 371–381.

Treisman, J.E., Heberlein, U. 1998. Eye development in *Drosophila*: Formation of the eye field and control of differentiation. Curr. Top. Dev. Biol. 39, 119–158.

Usui, T., Shima, Y., Shimada, Y., Hirano, S., Burgess, R.W., Schwarz, T.L., Takeichi, M., Uemura, T. 1999. Flamingo, a seven-pass transmembrane cadherin, regulates planar cell polarity under the control of *frizzled*. Cell 98, 585–595.

Van Buskirk, C., Schupbach, T. 1999. Versatility in *signaling*: Multiple responses to EGF receptor activation during *Drosophila* oogenesis. Trends Cell. Biol. 9, 1–4.

Veeman, M.T., Slusarski, D.C., Kaykas, A., Louie, S.H., Moon, R.T. 2003. Zebrafish prickle, a modulator of noncanonical Wnt/*fz* signaling, regulates gastrulation movements. Curr. Biol. 13, 680–685.

Vinson, C.R., Adler, P.N. 1987. Directional non-cell autonomy and the transmission of polarity information by the *frizzled* gene of *Drosophila*. Nature 329, 549–551.

Vinson, C.R., Conover, S., Adler, P.N. 1989. A *Drosophila* tissue polarity locus encodes a protein containing seven potential transmembrane domains. Nature 338, 263–264.

Wallingford, J.B., Goto, T., Keller, R., Harland, R.M. 2002. Cloning and expression of Xenopus Prickle, an orthologue of a *Drosophila* planar cell polarity gene. Mech. Dev. 116, 183–186.

Wang, W.-J., Liu, J., Adler, P.N. 1994. The *frizzled* gene of *Drosophila* encodes a membrane protein with an odd number of transmembrane domains. Mech. Dev. 45, 127–137.

Weber, U., Paricio, N., Mlodzik, M. 2000. Jun mediates *frizzled* induced R3/R4 cell fate distinction and planar polarity determination in the *Drosophila* eye. Development 127, 3619–3629.

Wehrli, M., Tomlinson, A. 1995. Epithelial planar polarity in the developing *Drosophila* eye. Development 121, 2451–2459.

Wehrli, M., Tomlinson, A. 1998. Independent regulation of anterior/posterior and equatorial/polar polarity in the *Drosophila* eye; evidence for the involvement of Wnt signaling in the equatorial/polar axis. Development 125, 1421–1432.

Winter, C.G., Wang, B., Ballew, A., Royou, A., Karess, R., Axelrod, J.D., Luo, L. 2001. *Drosophila* Rho-associated kinase (Drok) links *frizzled*-mediated planar cell polarity signaling to the actin cytoskeleton. Cell 105, 81–91.

Wolff, T., Ready, D.F. 1991. The beginning of pattern formation in the *Drosophila* compound eye: The morphogenetic furrow and the second mitotic wave. Development 113, 841–850.

Wolff, T., Ready, D.F. 1993. Pattern formation in *Drosophila* retina. In: *The development of* Drosophila *melanogaster* (M.B.A. Martinez-Arias, Ed.), Cold Spring Harbor: Cold Spring Harbor Press, pp. 1277–1326.

Wolff, T., Rubin, G.M. 1998. *strabismus*, a novel gene that regulates tissue polarity and cell fate decisions in *Drosophila*. Development 125, 1149–1159.

Wong, H.-C., Bourdelas, A., Krauss, A., Lee, H.-J., Shao, Y., Wu, D., Mlodzik, M., Shi, D.-L., Zheng, J. 2003. Direct binding of the PDZ domain of Disheveled to a conserved internal sequence in the C-terminal region of *frizzled*. Mol. Cell 12, 1251–1260.

Xu, Q., Wang, Y., Dabdoub, A., Smallwood, P.M., Williams, J., Woods, C., Kelley, M.W., Jiang, L., Tasman, W., Zhang, K., Nathans, J. 2004. Vascular development in the retina and inner ear: Control by Norrin and *frizzled*-4, a high-affinity ligand-receptor pair. Cell 116, 883–895.

Yanagawa, S., van Leeuwen, F., Wodarz, A., Klingensmith, J., Nusse, R. 1995. The Disheveled protein is modified by Wingless *signaling* in *Drosophila*. Genes Dev. 9, 1087–1097.

Yang, C., Axelrod, J.D., Simon, M.A. 2002. Regulation of *frizzled* by fat-like Cadherins during Planar Polarity Signaling in the *Drosophila* Compound Eye. Cell 108, 675–688.

Zeidler, M.P., Perrimon, N., Strutt, D.I. 1999. The four-jointed gene is required in the *Drosophila* eye for ommatidial polarity specification. Curr. Biol. 9, 1363–1372.

Zeidler, M.P., Perrimon, N., Strutt, D.I. 2000. Multiple roles for four-jointed in planar polarity and limb patterning. Dev. Biol. 228, 181–196.

Zheng, L., Zhang, J., Carthew, R.W. 1995. *frizzled* regulates mirror-symmetric pattern formation in the *Drosophila* eye. Development 121, 3045–3055.

Zipursky, S.L., Rubin, G.M. 1994. Determination of neuronal cell fate: lessons from the R7 neuron of *Drosophila*. Annu. Rev. Neurosci. 17, 373–397.

Long-range coordination of planar polarity patterning in *Drosophila*

Helen Strutt and David Strutt

Centre for Developmental Genetics, Department of Biomedical Science, University of Sheffield, Western Bank, Sheffield, S10 2TN, United Kingdom

Contents

One of the key questions regarding the phenomenon of planar polarity is how it is coordinated across large distances in a developing tissue or organ. To take a specific example, how is it that in the *Drosophila* wing, each one of the thousands of hairs comes to be arrayed precisely parallel to its neighbors and orientated towards the distal end of the wing? In large part, our understanding of the likely mechanisms underlying this long-range coordination of pattern has been guided by models based on the results of early transplantation experiments carried out in insects other than *Drosophila*. We will briefly describe some of these experiments and the conclusions drawn from them before discussing the molecular nature of signals controlling polarity patterning in *Drosophila* itself.

Advances in Developmental Biology
Volume 14 ISSN 1574-3349
DOI: 10.1016/S1574-3349(04)14003-9

1. Gradient models in the insect cuticle

The presence of directional polarity patterning information within arthropod segments was first demonstrated by transplantation experiments on the cuticles of a wide variety of insects. Particularly informative were studies by Locke on the blood-sucking insect *Rhodnius* (Locke, 1959). On the adult cuticle of *Rhodnius*, a simple pattern of ripples can be observed, which run laterally. If squares of larval cuticle are transplanted between equivalent anterior–posterior (AP) positions in different segments, the final cuticle pattern of the adult is unaffected. Interestingly, if transplantation is carried out between squares at different AP positions within the same segment, a reproducible whorled pattern of ripples is subsequently seen in the adult cuticle. A different pattern of whorls is observed if a square of tissue is rotated 180° (Fig. 1A). This led to the conclusion that each segment contains patterning information in an axial gradient, leading to cells in different positions in the gradient being incompatible in some way. This incompatibility was reflected in the orientation of ripples at the graft-host boundary—the "discontinuity pattern."

These results were subsequently interpreted in terms of the axial gradient being a gradient of a diffusible substance which runs between the boundaries of each segment (Lawrence, 1966; Stumpf, 1966). Thus, the ripples run perpendicular to the slope of the gradient, like contours on a map. If different values of this substance are apposed by transplantation, then the diffusible substance will flow from cells with high levels

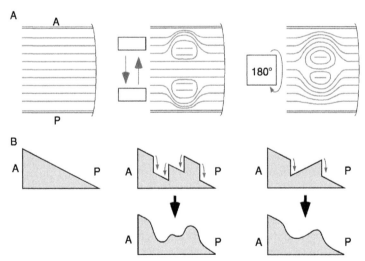

Fig. 1. Transplantation experiments in *Rhodnius*. (A) Pattern of ripples observed in each segment of the adult cuticle of Rhodnius (left). The ripples form whorled patterns if larval cuticle has been transplanted between anterior and posterior positions within a segment (middle) or rotated through 180° (right). (B) Gradient models to explain the whorled patterns in panel A: note that the AP axis is rotated 90° relative to (A). Discontinuities in an AP gradient occur after transplantation (middle) or rotation (right) of tissue. Flow of substance from high to low levels then occurs to give a smooth gradient profile with peaks and troughs.

to those with low levels, until equilibrium is achieved. The flow of substance will result in changes in the direction of the gradient (Fig. 1B), forming "valleys" and "hills" on a contour map, and this in turn causes the ripples to form circular whorls. Also consistent with this model are the results of rotating the entire middle section of a segment in the wax moth *Galleria* (Piepho, 1955); this causes scales in the central domain to be inverted, and those at the boundaries to be intermediate in orientation.

This gradient model was further refined to take account of the fact that the polarity patterns observed after transplantation are relatively stable over time. If the gradient merely consisted of a diffusible substance, then diffusion would rapidly abolish differences in levels (Lawrence et al., 1972). This led to the idea that the diffusible substance is not passively diffusing between the segment boundaries, but that each cell within the segment must contribute towards actively maintaining the gradient.

Subsequently a "differential adhesiveness" model was proposed, based on the reorientation of scales after transplantation in the wing of the tobacco hornworm *Manduca* (Nardi and Kafatos, 1976a,b). This differs from earlier models in that the axial gradient is not of a diffusible substance, but is a gradient of cell adhesion. Thus, in a graft of proximal tissue to a more distal position, the most adhesive cells in the graft (proximal cells) would tend to minimize their contacts with their neighbors and move towards the graft interior. Consequently, the more distal cells of the graft would tend to occupy more of the graft periphery. The "rosettes" of scales observed after such transplantation experiments are consistent with mathematical modeling of this process, assuming that cells move as a coherent layer. Furthermore, regeneration experiments in the milkweed bug *Oncopeltus* indicate that an abrupt change in cell affinities exists at the segment boundary. As segments are reiterating units, this supports the idea that there may be gradual changes in cell affinity throughout the entire segment (Wright and Lawrence, 1981).

Finally, it has also been observed that the cytoskeletons of cells in many tissues are closely aligned, due to intercellular contacts between cytoskeletal and plasma membrane proteins (Tucker, 1981). Modulation of such cell contacts may allow cells to transmit positional information along the axes of a tissue and be involved in establishing cell polarity.

It should be noted that whilst gradient models brought forward important concepts in polarity patterning, the studies were difficult to pursue further, as at that time tools were not available to identify the molecular basis of the gradient itself. More recently however, a number of molecules have been identified which may play a role in long-range polarity patterning, allowing the "pre-molecular" models to be translated into molecular terms.

2. Read-outs of planar polarity patterning in *Drosophila*

The wing, eye, and abdomen are the tissues of *Drosophila* which have been most studied for evidence of long-range polarity patterning mechanisms. In the wing, each cell produces a single hair which emerges from the distal vertex of the membrane

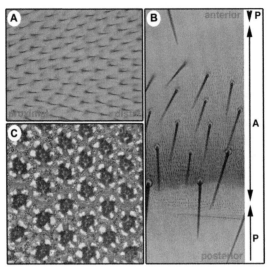

Fig. 2. Planar polarity in adult *Drosophila* tissues. (A) Each cell of the adult wing produces a single hair that points distally. (B) Hairs and bristles in the abdomen point posteriorly. Anterior (A) and posterior (P) compartments are shown. (C) Ommatidia have opposite polarity on either side of the DV midline, or equator (red line). (See Color Insert.)

(Fig. 2A). Similarly, the epidermal cells in the abdomen produce hairs and bristles which point to the posterior of the segment (Fig. 2B). In the eye, polarity is manifested at the level of the multicellular ommatidial units: Ommatidia in the dorsal half of the eye point in the opposite direction and have opposite chiralty to those in the ventral half of the eye (Fig. 2C). The mechanisms that coordinate long-range planar polarity patterning must ultimately act to direct the polarization of these diverse structures, most plausibly by employing a common cellular machinery. At least in the wing and eye, it is known that one of the earliest overt signs of planar polarity patterning involves the assembly of asymmetric protein complexes at the adherens junction zone of the cells (reviewed in Strutt, 2003). Thus, in the wing the seven-pass transmembrane protein *frizzled* (*fz*) and the cytoplasmic protein *disheveled* (*dsh*) are found on distal cell membranes (Axelrod, 2001; Shimada et al., 2001; Strutt, 2001), whilst the four-pass transmembrane protein *strabismus* (*stbm*) and the cytoplasmic protein *prickle-spiny-legs* (*pk-sple*) are found on proximal membranes (Tree et al., 2002; Bastock et al., 2003). The atypical cadherin *flamingo* (*fmi*) and the ankyrin-repeat protein *diego* are thought to be on both proximal and distal membranes (Usui et al., 1999; Feiguin et al., 2001). A similar asymmetric distribution is seen in the eye, with *fmi* on both sides and *fz/dsh* and *stbm/pk* on opposing sides of the boundary between the R3 and R4 photoreceptor cells (Das et al., 2002; Strutt et al., 2002; Yang et al., 2002; Jenny et al., 2003; Rawls and Wolff, 2003).

3. Gradients of secreted molecules in polarity patterning

In the eye and abdomen there is evidence for long-range gradients of secreted molecules, which act over the entire axis of the tissue in the eye, or in each segment of the abdomen. In addition to other roles in controlling cell fate specification, these molecules are required for correct polarity establishment. For instance, the Wingless (Wg) ligand is expressed at both the dorsal and ventral poles of the eye. Interestingly, clones of cells mutant for components of the Wg signal transduction pathway cause wild-type ommatidia on the boundary of the clone closest to the dorso-ventral (DV) midline (the equatorial boundary) to adopt inverted polarity on their DV axis (Wehrli and Tomlinson, 1998). Conversely, the Unpaired (Upd) ligand is expressed at the DV midline, and forms a gradient which is visible in the early second instar disc. Loss of function clones of components of the JAK/STAT signaling pathway also lead to nonautonomous inversions of ommatidial polarity. In this case, however, inversions are seen on boundaries that lie closer to the dorsal or ventral poles of the eye (the polar boundary) (Zeidler et al., 1999b). These phenotypes led to the conclusion that Wg and JAK/STAT signaling control the production of a gradient of another nonautonomously acting factor, and the orientation of this gradient directs ommatidial polarity (See Strutt and Strutt, 1999 for review). Clones mutant for genes in either pathway would cause a reversal of the direction of the gradient on one side of the clone (Fig. 3A).

In the abdomen, Hedgehog (Hh) protein is made in the posterior compartment of each segment, and diffuses into the neighboring anterior compartments to form a concentration gradient. Clones of cells overexpressing *hh* in the anterior

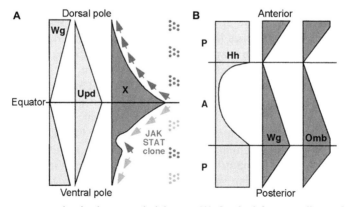

Fig. 3. Long-range patterning in the eye and abdomen. (A) On the left are gradients of Wg and Upd ligands in the eye imaginal disc (light grey). On the right is a gradient of a putative secondary signaling molecule X (dark grey) which is regulated by Wg and JAK/STAT signaling; the direction of the gradient controls ommatidial orientation. A clone of JAK-STAT signaling component alters the direction of the gradient on the polar side of the clone. (B) In the abdomen, Hh protein produced in the P compartment diffuses into the A compartment (left), which regulates the expression of Wg and *omb* in gradients (right).

compartment cause wild-type hairs posterior to the clone to point towards the clone. Therefore, *hh* was proposed to control the production of a secondary signaling molecule that regulates polarity (Struhl et al., 1997a,b; Lawrence et al., 1999a,b). More recently, it was proposed that this activity of *hh* is mediated by the induction of *wg* and *optomotor-blind (omb)* expression at the AP boundary (Lawrence et al., 2002) (Fig. 3B).

Equivalent secreted signaling molecules that pattern the proximo-distal (PD) axis of the wing have not yet been identified. Interestingly, Wg is again a good candidate for fulfilling this function, as it is expressed on the DV boundary of the wing imaginal disc. However direct evidence for a role of Wg signaling in PD planar polarity patterning is lacking at present.

Each of the long-range patterning systems described above is specific to the tissue in question. However, several genes have been identified that are involved in regulating transmission of polarity patterning information in multiple tissues, and act downstream of long-range patterning cues. We will first describe how the *four-jointed (fj)*, *dachsous (ds)*, and *fat (ft)* genes may act in this process. The role of some of the genes involved in asymmetric complex formation, namely *fz* and *stbm*, in long-range polarity patterning will then be discussed, before addressing how the two groups of genes might act together.

4. *four-jointed*, *dachsous*, and *fat* phenotypes in polarity patterning

fj, *ds*, and *ft* were first studied because mutants had defects in PD patterning of the limbs. The wings are shorter than normal and the leg segments are also reduced in length, as well as some leg segments being fused together (Mohr, 1923; Stern and Bridges, 1926; Waddington, 1943; Mahoney et al., 1991; Clark et al., 1995; Villano and Katz, 1995; Brodsky and Steller, 1996). Genetic interactions have been demonstrated between all three genes (Mohr, 1929; Villano and Katz, 1995; Strutt et al., 2004), leading to speculation that they act in a common PD patterning process.

fj encodes a type II transmembrane protein, which can be cleaved near its transmembrane domain (Villano and Katz, 1995; Brodsky and Steller, 1996; Buckles et al., 2001). This causes the release of the extracellular, C-terminal region of the protein, and it was therefore proposed that *fj* acts as a secreted signaling molecule. However, recent data suggest that *fj* is found predominantly in the Golgi, and secretion of the C-terminal region is not essential for *fj* function (Strutt et al., 2004). Nevertheless, *fj* has been shown to act nonautonomously (Tokunaga and Gerhart, 1976; Zeidler et al., 1999a, 2000), indicating that it must somehow modify the activity of one or more other molecules involved in intercellular communication.

ds and *ft* both encode atypical cadherins; their products are transmembrane proteins with very large extracellular domains consisting of 27 and 34 tandem cadherin repeats, respectively. *ft* also contains 4 EGF-like repeats and *LamininA* G-domain-type repeats in its extracellular domain (Mahoney et al., 1991; Clark et al., 1995).

It has been shown that opposing gradients of expression or activity of *fj*, *ds*, and *ft* are present in the eye, wing, and abdomen, that regulate planar polarity patterning.

They have been most studied in the eye, where *fj* is expressed in a gradient, with high levels at the DV midline and low levels at the polar boundaries (Villano and Katz, 1995; Brodsky and Steller, 1996). This expression is positively regulated by JAK/STAT signaling and negatively regulated by Wg signaling (Zeidler et al., 1999a). *ds* is, however, positively regulated by Wg signaling, and is expressed at high levels at the poles of the eye (Clark et al., 1995; Yang et al., 2002), whilst *ft* is uniformly expressed (Mahoney et al., 1991; Yang et al., 2002). Interestingly, *fj* homozygous mutants have only very minor polarity defects, but clones of cells mutant for *fj* cause nonautonomous inversions of ommatidial polarity on the polar boundary of the clone (Fig. 4A) (Zeidler et al., 1999a). *ds* and *ft* mutants show a slightly different phenotype: In homozygotes and throughout loss of function clones, ommatidia have randomized DV polarity. In addition, wild-type ommatidia on the equatorial boundary of *ds* clones are often inverted (Fig. 4B), whereas in *ft* clones, inversions are found

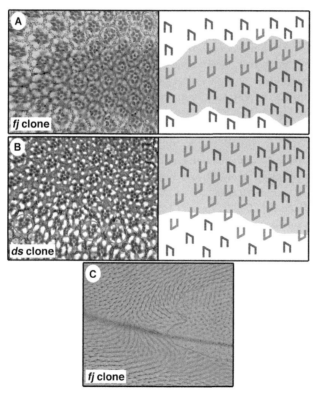

Fig. 4. *fj* and *ds* mutant phenotypes. (A,B) *fj* eye clone (A) and *ds* eye clone (B) marked by loss of pigment (left) or grey shading (right). In the left panels, wild-type ommatidia with reversed DV polarity are circled, whilst in the right panels all ommatidia with reversed polarity are shown in red. Note that several rows of ommatidia on the polar side of the *fj* clone have reversed orientation (A); whilst inside the *ds* clone ommatidia have randomized polarity and nonautonomy is on the equatorial boundary (B). (C) *fj* wing clone, mutant tissue marked by the hair morphology marker *shavenoid*, and outlined in red. Wild-type hairs proximal to the clone point away from the clone. (See Color Insert.)

on the polar boundary (Rawls et al., 2002; Strutt and Strutt, 2002; Yang et al., 2002; Fanto et al., 2003). Therefore, *fj* signals in the same direction as *ft* in polarity patterning, but *ds* has opposite signaling activity (see Table 1).

Epistasis experiments subsequently showed that *ds* acts downstream of *fj* (Strutt and Strutt, 2002; Yang et al., 2002). The position of *ft* in this hierarchy is less clear, as epistasis experiments between *ds* and *ft* did not yield a conclusive result (Yang et al., 2002).

Similar opposing gradients of activity have been observed in the abdomen and wing. In the abdomen, *fj-lacZ* is expressed in a stripe in the anterior region of the anterior compartment (Zeidler et al., 2000; Casal et al., 2002), and its expression is postulated to be regulated by *hh* or *omb*. Similarly, *ds-lacZ* is expressed in a stripe within each segment, extending into both compartments with highest levels at the AP boundary (Casal et al., 2002). Clones of cells mutant for *fj* in the anterior compartment cause wild-type hairs posterior to the clone to reverse their polarity; although they have no phenotype in the posterior compartment (Zeidler et al., 2000; Casal et al., 2002). Analysis of *ds* and *ft* clones in the anterior compartment showed that, as in the eye, *ft* nonautonomy acts in the same direction as *fj*, whilst *ds* acts oppositely (Casal et al., 2002). Unlike *fj*, both *ds* and *ft* clones also have nonautonomous phenotypes in the posterior compartment, however they act in the opposite direction to clones in the anterior compartment (Table 1). These phenotypes suggest that *ds* and *ft* have opposite activity gradients in each compartment of the abdomen.

Finally, *fj* protein has a graded distribution in the wing imaginal disc, with highest levels at the center of the wing pouch, which will go on to be the distal end of the wing (Strutt et al., 2004). By pupal stages, a *fj-lacZ* enhancer trap is expressed at highest levels at the distal end of the wing (Zeidler et al., 2000; Ma et al., 2003). *ds* is expressed most strongly in the future hinge region of the wing imaginal disc and in the hinge and proximal wing region of the pupal wing (Clark et al., 1995; Strutt and Strutt, 2002; Ma et al., 2003). Clones of cells mutant for *fj* or *ft* affect the polarity of wild-type hairs on the proximal side of the clone, causing them to point away from the clone (Fig. 4C) (Zeidler et al., 2000; Strutt and Strutt, 2002; Ma et al., 2003). In contrast, *ds* clones show distal nonautonomy (Table 1) (Adler et al., 1998). Interestingly, there is a gradient of phenotypic strength for both *fj* and *ft* clones, such that clones in the center of the wing have the strongest nonautonomous effects (Zeidler et al., 2000; Strutt and Strutt, 2002).

5. The function of *four-jointed*, *dachsous*, and *fat* in polarity patterning

A number of lines of evidence suggest that *fj*, *ds*, and *ft* act upstream of overt signs of polarity patterning such as asymmetric protein localization. Firstly, in *ds* homozygotes, the wing hairs form a distinctive swirling pattern, which is different from that seen for mutants in any of the genes involved in asymmetric complex formation, such as *fz* and *dsh* (Adler et al., 1998). Notably, in flies doubly mutant for *fz* and *ds*, the wings have a *fz*-like hair pattern, suggesting that *fz* is acting downstream of *ds* (Adler et al., 1998). In addition, in pupal wings of *fj*, *ds*, and *ft* mutants, asymmetric protein

Table 1
Phenotypes of genes involved in long-range patterning

	fj	*ds*	*ft*	*fz*	*stbm*
Eye					
Homozygote	Low frequency of DV inversions	DV inversions	DV inversions	DV inversions and misrotations[a]	DV inversions and misrotations[a]
Clone boundary	DV inversions (polar boundary)	DV inversions (equatorial boundary)	DV inversions (polar boundary)	DV inversions (polar boundary)	No phenotype
Abdomen					
Homozygote	No phenotype	Bristles and hairs form swirling patterns	No data	Bristles and hairs form swirling patterns[a]	Bristles and hairs form swirling patterns[a]
Clone boundary - A compartment	Hairs point towards clone (posterior to clone)	Hairs point away from clone (anterior to clone)	Hairs point towards clone (posterior to clone)	Hairs point towards clone (posterior to clone)	Hairs point away from clone (anterior to clone)
Clone boundary - P compartment	No phenotype	Hairs point towards clone (posterior to clone)	Hairs point away from clone (anterior to clone)	Hairs point towards clone (posterior to clone)	No data
Wing					
Homozygote	No phenotype	Hairs form polarity swirls[b]	No data	Hairs form polarity swirls[a]	Hairs form polarity swirls[a]
Clone boundary	Hairs point towards clone (distal to clone)	Hairs point away from clone (proximal to clone)	Hairs point towards clone (distal to clone)	Hairs point towards clone (distal to clone)	Hairs point away from clone (proximal to clone)

[a]Phenotypes due to cell autonomous patterning activity.
[b]Pattern of swirls distinct from the *fz/stbm* pattern.

complexes still form, but at abnormal positions at the cell periphery. This leads to emergence of the prehair at the same abnormal site (Adler et al., 1998; Strutt and Strutt, 2002; Ma et al., 2003). Similarly, in the eyes of *ds* and *ft* mutants, asymmetric protein complexes form in the R3 and R4 photoreceptor cells, but they often have reversed orientation (Strutt and Strutt, 2002; Yang et al., 2002). Therefore it is thought that *fj*, *ds*, and *ft* must act in the transmission of a polarity signal that occurs before asymmetric complex formation. Mutations in these genes do not disrupt formation of asymmetric complexes *per se*, but they lead to complex formation in an aberrant pattern.

As *ds* and *ft* proteins have cadherin repeats in their extracellular domains, it is possible that homophilic and/or heterophilic interactions between them are important for polarity patterning (Mahoney et al., 1991; Clark et al., 1995). Indeed, clones of cells mutant for *ds* tend to be rounded up and bulge from the wing surface, suggesting that *ds* has a role in cell adhesion (Adler et al., 1998). It has been shown that *ft* protein is present at membranes of wild-type cells adjoining *ft* clones, and *ds* protein at membranes of cells adjacent to *ds* clones, implying that neither *ds* nor *ft* have to interact homophilically (Ma et al., 2003). However, in *ft* clones *ds* staining is diffuse within the cell, and no longer tightly associated with the plasma membrane; in addition, *ds* staining is stronger than normal on the cell membranes between ft^+ and ft^- tissue. A similar effect is seen for *ft* protein in *ds* clones (Strutt and Strutt, 2002; Ma et al., 2003): this would be consistent with the idea that heterophilic interactions between *ds* and *ft* are necessary for correct membrane localization.

It has also been shown that *fj* can affect membrane localization of *ds* and *ft*. A boundary effect can be observed in *fj* pupal wing clones, such that in the row of mutant cells on the edge of the clone, both *ds* and *ft* are concentrated on the membranes between fj^+ and fj^- tissue, and staining is lost on membranes between mutant cells (Strutt and Strutt, 2002; Ma et al., 2003). In addition, clones of *fj* in the wing imaginal disc tend to round up, and cells inside the clone are more tightly packed than normal (Strutt and Strutt, 2002). Therefore, one role of *fj* could be to modify the properties of *ds* and *ft*, leading to changes of adhesion between cells.

These data support models in which *fj* might regulate a gradient of cell adhesion along the PD axis of the wing, the DV axis of the eye and the AP axis of each compartment of the abdomen; and this gradient would act to transmit polarity information. However, it has recently been shown that *ft* binds to the transcriptional co-repressor Atrophin (Atro) (Fanto et al., 2003). Clones of cells mutant for *atro* have similar polarity phenotypes to *ft* in the eye and wing (Zhang et al., 2002; Fanto et al., 2003), and clones have smooth borders, suggesting that *atro* is also required for cell adhesion. It is not clear why an Atro-mediated downstream transcriptional response would be necessary, if the only mechanism of *ds* and *ft* function in planar polarity patterning was to produce a gradient of cell adhesion via heterophilic interactions. Another possibility therefore is that *fj/ds/ft/atro* act to regulate the production of a short range signaling molecule, in proportion to the amount of *atro* activity: this would lead to a gradient of positional information that directs cell polarization. Alternatively, interactions between *ds* and *ft* could be required for the directional relay of information from cell to cell.

An interesting problem is that polarity phenotypes are only observed in *fj* mutant clones, whilst *fj* homozygous animals do not show significant polarity phenotypes (Zeidler et al., 1999a, 2000). In contrast, polarity phenotypes are seen in tissue wholly mutant for *ds* and *ft* (Adler et al., 1998; Rawls et al., 2002; Strutt and Strutt, 2002; Yang et al., 2002). The lack of polarity phenotypes in *fj* mutants has been explained by proposing that it acts redundantly to regulate the production of a gradient—therefore loss of *fj* throughout the tissue would not affect the direction of the gradient, but a loss of function clone would cause a discontinuity in the gradient on one side of the clone, leading to inversions of polarity. This redundancy suggests that *fj* may modulate *ds* and/or *ft* activity, but they are also regulated by other inputs, such that they retain some activity in the absence of *fj*. For example, in the eye *wg* positively regulates *ds* expression in a gradient (Yang et al., 2002); *fj* may then act on top of this to further modulate the overall activity of *ds*.

A related point is that *fj* and *ft* only show strong nonautonomous polarity phenotypes in the center of the wing (Zeidler et al., 2000; Strutt and Strutt, 2002). This is far from the distal region of the wing, where *fj* and *ft* are most highly expressed. The reason for the variation in phenotypic strength is not clear, but it is possible that the strong phenotypes indicate the region of the wing where the activity gradients of *fj* and *ft* are steepest. Support for this comes from the observation that PD patterning defects in *fj* mutants are also more pronounced in the same central region of the wing, laying between the anterior and posterior cross-veins (Villano and Katz, 1995; Brodsky and Steller, 1996). Hence a graded distribution of *fj/ds/ft* activity could control both PD patterning and polarity patterning. Whether these processes are linked or if they occur independently is an as yet unanswered question.

6. Nonautonomous signaling activity of *frizzled* and *strabismus*

The *fz* transmembrane protein is a component of the asymmetric protein complexes which direct wing hair polarity and ommatidial polarity (Strutt, 2001; Strutt et al., 2002). Interestingly, clones of cells mutant for *fz* also show nonautonomous effects on planar polarity patterning in the wing, eye, and abdomen. In the wing, wild-type hairs on the distal side of the clone tend to reverse their polarity and point towards the clone (Fig. 5A and Table 1) (Gubb and García-Bellido, 1982; Vinson and Adler, 1987); this is accompanied by reorientation of polarity protein complexes at the prehair initiation site (Usui et al., 1999). Similarly, in the abdomen, hairs posterior to a *fz* clone point towards the clone (Casal et al., 2002; Lee and Adler, 2002). In *fz* homozygous eyes and inside *fz* eye clones, the ommatidia adopt a random polarity characteristic of loss of asymmetric complexes; however on the polar boundary of the clone a number of wild-type ommatidia become inverted on the DV axis (Fig. 5B) (Zheng et al., 1995).

Similar nonautonomy has been observed for *stbm* in the wing and abdomen, although in each case the direction of nonautonomy is the opposite to that of *fz* (Table 1) (Taylor et al., 1998; Lee and Adler, 2002). *stbm* has, however, been reported not to show nonautonomous phenotypes in the eye (Wolff and Rubin, 1998). Clones

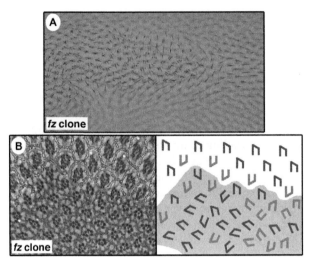

Fig. 5. *fz* eye and wing phenotypes. (A) *fz* clone in the wing, mutant tissue marked by the *multiple wing hairs* gene, and outlined in red. Hairs distally and laterally to the clone point towards the clone. (B) *fz* eye clone, marked by loss of pigment (left) or grey shading (right). Inside the clone ommatidia have randomized polarity and orientation, and some wild-type ommatidia on the polar clone boundary have reversed DV polarity (circled in red). Dorsal-type ommatidia are in blue and ventral-type in red. (See Color Insert.)

of *pk-sple* have nonautonomous phenotypes that are less extensive than either *fz* or *stbm* (Gubb and García-Bellido, 1982; Gubb et al., 1999); whereas, the other genes involved in asymmetric complex formation, *fmi* and *dsh*, do not show any significant nonautonomous activity (Klingensmith et al., 1994; Theisen et al., 1994; Chae et al., 1999; Usui et al., 1999; Strutt and Strutt, 2002). Therefore, there is a strong requirement for both *fz* and *stbm* and a weak requirement for *pk-sple*, for the transmission of polarity information along the PD axis of the wing, in addition to *fj*, *ds*, and *ft* function.

If *fz* is overexpressed at the distal end of the wing, hair polarity is reversed, such that hairs point away from the distal end of the wing (Adler et al., 1997). Similarly, clones of cells overexpressing *fz* cause hairs proximally to the clone to point away from the clone (Krasnow and Adler, 1994; Strutt, 2001). Together with the loss of function phenotype of *fz*, where hairs point towards the mutant tissue, these results suggest that hairs point away from high levels of *fz* activity and towards low levels of *fz* activity. In contrast, hairs point towards high levels of *stbm* activity (Adler et al., 2000).

Two model types have been proposed to explain the dual function of *fz* in polarity patterning, such that it acts both autonomously and nonautonomously. In cell–cell relay models, *fz* acts by receiving a polarizing ligand at the proximal end of the cell. This leads to the activation of two signal transduction pathways. One pathway would regulate placement of the hair at the distal end of the cell; the other pathway would release more ligand to signal to neighboring distal cells (Park et al., 1994). In secondary signal models, a long range gradient is presumed to cause a gradient of *fz* activation. This in turn results in the proportional release of a locally-acting signal that then acts again via *frizzled* to polarize cells (Zheng et al., 1995; Adler et al., 1997).

The cell–cell relay model has recently been elaborated based on the asymmetric distribution of polarity protein complexes in the wing (Tree et al., 2002). Interactions between asymmetric complexes are envisaged between the distal membrane of one cell and the proximal membrane of the neighboring cell. Loss of asymmetric complexes within a clone may therefore cause polarity complexes in the immediate neighbors to have reversed orientation; and this could be propagated from cell to cell. It has also been suggested that propagation of polarity complexes could be a mechanism for the precise alignment of cells within a tissue, acting to ensure fidelity in the reading of global signals regulated by *fj*, *ds*, and *ft* (Ma et al., 2003).

A number of observations argue against propagation of asymmetric complexes being the major mechanism for *fz/stbm*-dependent transmission of polarity information. Firstly, of the proteins known to be required for asymmetric complexes to form, only *fz* and *stbm*, and to a lesser extent *pk-sple*, act nonautonomously. In addition, alleles of *fz* have been identified which only disrupt its cell autonomous activity. In these alleles asymmetric complexes fail to form, but nonautonomous signaling is normal (Strutt, 2001). Similarly, both *fmi* and *dsh* mutants disrupt the formation of polarity complexes, but do not affect propagation of polarity information to neighboring cells. (But note that *fmi* can, however, act nonautonomously if overexpressed (Feiguin et al., 2001)). In addition, cell-to-cell propagation of complexes cannot explain *fz* nonautonomy in the eye, where other cell types separate the R3 and R4 photoreceptors that localize polarity proteins (as noted by Fanto et al., 2003).

Finally, it has recently been demonstrated that the cell autonomous and nonautonomous functions of *fz* are temporally separable (Strutt and Strutt, 2002). Rescue experiments were carried out by expressing a *fz* transgene at defined times of development. In both the wing and eye the formation of asymmetric complexes can be rescued significantly later than nonautonomous signaling activity, suggesting that nonautonomous signaling precedes autonomous activity and localization of polarity complexes. Consistent with this, no nonautonomy is seen if *fz* or *stbm* are overexpressed, or loss of function clones are induced, in a background homozygous mutant for any of the other genes involved in complex formation (Adler et al., 1997; Taylor et al., 1998; Chae et al., 1999; Lee and Adler, 2002). This indicates that autonomous planar polarity patterning is epistatic to nonautonomous activity and thus is a downstream event.

Therefore, it seems likely that if there is a cell–cell relay mechanism which depends on sequential assembly of asymmetric protein complexes from cell to cell, at best this is probably a short-range system which acts relatively late to refine patterning. Gradient models, or cell–cell relay models that do not involve propagation of complexes, are equally plausible alternatives to explain the nonautonomous behavior of *fz* and *stbm*, and indeed would be more consistent with the experimental observations. However it should be noted, that no ligands have been identified that would cause a *fz* or *stbm* activity gradient.

7. Interaction between the *fj*/*ds*/*ft* and *fz*/*stbm* pathways

The data so far indicate that two systems are involved in regulating the propagation of polarity information across developing epithelia: firstly, gradients of *fj*, *ds*, and *ft* activity and secondly, *fz* and *stbm* nonautonomous activity. The relationship between these two systems has been little studied thus far. Interestingly, if transgene rescue technology is used, such that only the later cell autonomous function of *fz* is present, the wing hairs form a swirling pattern that resembles the *ds* pattern (Taylor et al., 1998; Strutt and Strutt, 2002), suggesting that this pattern is characteristic of a specific loss of nonautonomous planar polarity patterning activity. Furthermore, in the eye, ommatidia do not take on a random orientation, but many are inverted on the DV axis, again similar to *ds* or *ft* eyes (Rawls et al., 2002; Strutt and Strutt, 2002; Yang et al., 2002).

A weak *fz* phenotype, or a *fz* nonautonomous phenotype can be enhanced by a weak *ds* mutation (Adler et al., 1998; Strutt and Strutt, 2002); furthermore, the extent of *fz* nonautonomy is increased in a *ds* or *ft* background (Adler et al., 1998; Ma et al., 2003). These results suggest that the loss of both pathways has an additive effect and that one pathway is not upstream of the other. Consistent with this, eye clones mutant for both *ds* and *fz* do not show epistasis, but result in an additive phenotype in which nonautonomy extends on both equatorial and polar boundaries of the clone (Strutt and Strutt, 2002). Therefore, the most likely hypothesis is that the two systems act as parallel pathways, and that transmission of polarity information is a combination of two systems feeding into the final read-out of polarity. However, it should be noted that these pathways are not redundant: tissue entirely homozygous for mutations in a single component of either pathway results in defects in transmission of polarity information. Therefore both pathways must function simultaneously to correctly establish long-range polarity patterning.

Very little is known about the exact time period in which either *fz*/*stbm* or *fj*/*ds*/*ft* nonautonomous activity is required. In the wing, rescue experiments suggest that a loss of *fz* nonautonomous activity can be completely rescued if a transgene is expressed before six hours after formation of the prepupa (Strutt and Strutt, 2002). Therefore endogenous *fz* must be functioning at this time, and possibly earlier. Similar experiments in the eye suggest that the requirement for *fz* nonautonomous activity begins before passage of the morphogenetic furrow and differentiation of photoreceptor cells (Strutt and Strutt, 2002). No timing experiments have been carried out for *fj*, *ds*, or *ft*. Overexpression of *fj* in the eye using photoreceptor cell-specific promoters can cause polarity defects, suggesting that *fj* may act late in development (Zeidler et al., 1999a). However, the ability to cause phenotypes when overexpressed does not mean that the endogenous protein acts at this time. Indeed, the graded distributions of *fj-lacZ* or *ds* protein are evident well ahead of the morphogenetic furrow (Villano and Katz, 1995; Brodsky and Steller, 1996; Yang et al., 2002), suggesting that they could act before photoreceptor differentiation, like *fz*. In the developing wing, it is notable that the most clear evidence for *fj* protein being in a gradient is in the third instar wing disc, whilst in the prepupal wing protein levels are lower, with protein at the distal tip (Strutt et al., 2004). Staining of a *fj-lacZ*

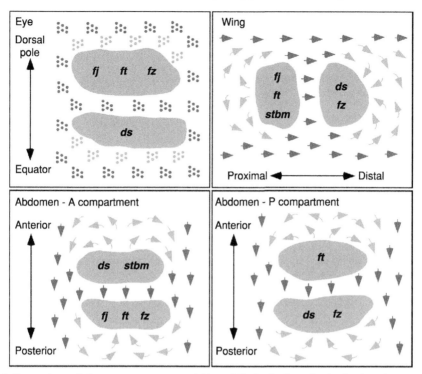

Fig. 6. Comparison of the nonautonomous activities of *fj*, *ds*, *ft*, *fz*, and *stbm* in the wing, eye, and abdomen. The position of loss of function clones for the genes indicated are shaded in grey (wild-type tissue is white). Wild-type ommatidia or hairs are in dark grey and nonautonomous disruptions of polarity on one side of the clone are in light grey.

enhancer trap however appears to show a gradient at later stages (Zeidler et al., 2000; Ma et al., 2003). Therefore, the requirements for both *fj* and *fz* activity could possibly be in the prepupal or even the third instar stage of development.

An interesting point to note is that whilst the two patterning systems appear to act in parallel, they do not always show nonautonomous phenotypes in the same relative direction (Fig. 6). For example, *fj* always shows nonautonomous phenotypes on the same side of clones as *ft*, whilst *ds* acts in the opposite direction (Zeidler et al., 1999a,b; Rawls et al., 2002; Strutt and Strutt, 2002; Yang et al., 2002). Similarly, *fz* and *stbm* nonautonomy is always seen on the opposite sides of clones (Vinson and Adler, 1987; Taylor et al., 1998; Lee and Adler, 2002). However, *fz* and *ft* show nonautonomy on the same side of eye clones (Zheng et al., 1995; Zeidler et al., 1999a), but the opposite side of wing clones (Vinson and Adler, 1987; Zeidler et al., 2000). The situation is even more complex in the abdomen: in the anterior compartment *fz* and *ft* signal in the same direction, as in the eye; whilst in the posterior compartment they signal in the opposite direction, as in the wing (Casal et al., 2002). This tells us that even though all tissues employ the same molecules to propagate

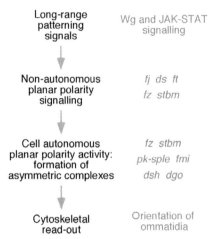

Fig. 7. Scheme of planar polarity patterning events. On the left is a general scheme for the conversion of long-range patterning signals into the final read-out of planar polarity, and on the right are some of the genes involved in this process in the eye.

polarity information, the way in which patterning information is translated into cell polarization requires tissue-specific responses.

To summarize, we have discussed evidence for the roles of two groups of genes, *fz* and *stbm*, and *fj*, *ds*, and *ft*, in regulating transmission of polarity information in the eye, wing, and abdomen of *Drosophila*. It is likely that the two systems act jointly to maintain the polarity information produced by short-lived gradients of long-range secreted signals, at least in the eye and abdomen. This then leads to the formation of asymmetric complexes in a uniform polarized manner (Fig. 7).

Despite the advances in identifying molecules involved in long-range polarity patterning, there is much that is not yet understood. The models developed in the pre-molecular studies on insect cuticles may still provide useful concepts for understanding the process of long-range patterning. For example, the phenotypes given by the identified genes are consistent with gradient models; and it would be attractive to suggest that the known proteins may act by regulating the secretion of short-range secondary signals. However, other mechanisms are also consistent with the data. Support for differential adhesivity models can be found at least for the *fj/ ds/ft* pathway, as the molecular structure of *ds* and *ft* suggests a role for cell adhesion affecting cell–cell communication. Alternatively, these molecules could be acting in a cell–cell relay. Elucidating the mechanisms of long-range patterning will require further study of the functions of these proteins and of their molecular interactions.

References

Adler, P., Charlton, J., Liu, J. 1998. Mutations in the cadherin superfamily member gene *dachsous* cause a tissue polarity phenotype by altering *frizzled* signaling. Development 125, 959–968.

Adler, P.N., Krasnow, R.E., Liu, J. 1997. Tissue polarity points from cells that have higher Frizzled levels towards cells that have lower Frizzled levels. Curr. Biol. 7, 940–949.

Adler, P.N., Taylor, J., Charlton, J. 2000. The domineering non-autonomy of *frizzled* and *Van Gogh* clones in the *Drosophila* wing is a consequence of a disruption in local signaling. Mech. Dev. 96, 197–207.

Axelrod, J.D. 2001. Unipolar membrane association of Dishevelled mediates Frizzled planar cell polarity signaling. Genes Dev. 15, 1182–1187.

Bastock, R., Strutt, H., Strutt, D. 2003. Strabismus is asymmetrically localised and binds to Prickle and Dishevelled during *Drosophila* planar polarity patterning. Development 130, 3007–3014.

Brodsky, M.H., Steller, H. 1996. Positional information along the dorsal-ventral axis of the *Drosophila* eye: Graded expression of the *four-jointed* gene. Dev. Biol. 173, 428–446.

Buckles, G.R., Rauskolb, C., Villano, J.L., Katz, F.N. 2001. *four-jointed* interacts with *dachs, abelson* and *enabled* and feeds back onto the *Notch* pathway to affect growth and segmentation in the *Drosophila* leg. Development 128, 3533–3542.

Casal, J., Struhl, G., Lawrence, P. 2002. Developmental compartments and planar polarity in *Drosophila*. Curr. Biol. 12, 1189–1198.

Chae, J., Kim, M.J., Goo, J.H., Collier, S., Gubb, D., Charlton, J., Adler, P.N., Park, W.J. 1999. The *Drosophila* tissue polarity gene *starry night* encodes a member of the protocadherin family. Development 126, 5421–5429.

Clark, H.F., Brentrup, D., Schneitz, K., Bieber, A., Goodman, C., Noll, M. 1995. *Dachsous* encodes a member of the cadherin superfamily that controls imaginal disc morphogenesis in *Drosophila*. Genes Dev. 9, 1530–1542.

Das, G., Reynolds-Kenneally, J., Mlodzik, M. 2002. The atypical cadherin Flamingo links Frizzled and Notch signaling in planar polarity establishment in the *Drosophila* eye. Dev. Cell 2, 655–666.

Fanto, M., Clayton, L., Meredith, J., Hardiman, K., Charroux, B., Kerridge, S., McNeill, H. 2003. The tumor-suppressor and cell adhesion molecule Fat controls planar polarity via physical interactions with Atrophin, a transcriptional co-repressor. Development 130, 763–774.

Feiguin, F., Hannus, M., Mlodzik, M., Eaton, S. 2001. The ankyrin-repeat protein Diego mediates Frizzled-dependent planar polarisation. Dev. Cell 1, 93–101.

Gubb, D., García-Bellido, A. 1982. A genetic analysis of the determination of cuticular polarity during development in *Drosophila melanogaster*. J. Embryol. Exp. Morphol. 68, 37–57.

Gubb, D., Green, C., Huen, D., Coulson, D., Johnson, G., Tree, D., Collier, S., Roote, J. 1999. The balance between isoforms of the Prickle LIM domain protein is critical for planar polarity in *Drosophila* imaginal discs. Genes Dev. 13, 2315–2327.

Jenny, A., Darken, R.S., Wilson, P.A., Mlodzik, M. 2003. Prickle and Strabismus form a functional complex to generate a correct axis during planar cell polarity signaling. EMBO J. 22, 4409–4420.

Klingensmith, J., Nusse, R., Perrimon, N. 1994. The *Drosophila* segment polarity gene *disheveled* encodes a novel protein required for response to the *wingless* signal. Genes Dev. 8, 118–130.

Krasnow, R.E., Adler, P.N. 1994. A single *frizzled* protein has a dual function in tissue polarity. Development 120, 1883–1893.

Lawrence, P.A. 1966. Gradients in the insect segment: The orientation of hairs in the milkweed bug *Oncopeltus fasciatus*. J. Exp. Biol. 44, 607–620.

Lawrence, P.A., Casal, J., Struhl, G. 1999a. hedgehog and engrailed: Pattern formation and polarity in the *Drosophila* abdomen. Development 126, 2431–2439.

Lawrence, P.A., Casal, J., Struhl, G. 1999b. The hedgehog morphogen and gradients of cell affinity in the abdomen of *Drosophila*. Development 126, 2441–2449.

Lawrence, P.A., Casal, J., Struhl, G. 2002. Towards a model of the organisation of planar polarity and pattern in the *Drosophila* abdomen. Development 129, 2749–2760.

Lawrence, P.A., Crick, F.H.C., Munro, M. 1972. A gradient of positional information in an insect, *Rhodnius*. J. Cell Sci. 11, 815–853.

Lee, H., Adler, P.N. 2002. The function of the *frizzled* pathway in the *Drosophila* wing is dependent on *inturned* and *fuzzy*. Genetics 160, 1535–1547.

Locke, M. 1959. The cuticular pattern in an insect, *Rhodnius Prolixus*. J. Exp. Biol. 36, 459–477.

Ma, D., Yang, C.H., McNeill, H., Simon, M.A., Axelrod, J.D. 2003. Fidelity in planar cell polarity signaling. Nature 421, 543–547.

Mahoney, P.A., Weber, U., Onofrechuk, P., Biessmann, H., Bryant, P.J., Goodman, C.S. 1991. The *fat* tumor suppressor gene in *Drosophila* encodes a novel member of the cadherin gene superfamily. Cell 67, 853–868.

Mohr, O. 1923. Modifications of the sex ratio through a sex-linked semi-lethal in *Drosophila melanogaster* (besides notes on an autosomal section deficiency). In: *Studia Mendeliana: Ad centesimum diem natalem Gregorii Mendelii a grata patria celebrandum,* Brünn, Czechoslovakia: Apud Typos, pp. 266–287.

Mohr, O. 1929. Exaggeration and inhibition phenomena encountered in the analysis of an autosomal dominant. Z. Indukt. Abstammungs-Vererbungsl. 50, 113–200.

Nardi, J.B., Kafatos, F.C. 1976a. Polarity and gradients in lepidopteran wing epidermis. I. Changes in graft polarity, form and cell density accompanying transpositions and reorientations. J. Embryol. exp. Morph. 36, 469–487.

Nardi, J.B., Kafatos, F.C. 1976b. Polarity and gradients in lepidopteran wing epidermis. II. The differential adhesiveness model: gradient of a non-diffusible cell surface parameter. J. Embryol. exp. Morph. 36, 489–512.

Park, W.J., Liu, J., Adler, P.N. 1994. The *frizzled* gene of *Drosophila* encodes a membrane protein with an odd number of transmembrane domains. Mech. Dev. 45, 127–137.

Piepho, H. 1955. Über die Ausrichtung der Schuppenbälge und Schuppen am Schmetterlingsrumpf. Naturwissenschaften 42, 22.

Rawls, A.S., Guinto, J.B., Wolff, T. 2002. The cadherins Fat and Dachsous regulate dorsal/ventral signaling in the *Drosophila* eye. Curr. Biol. 12, 1021–1026.

Rawls, A.S., Wolff, T. 2003. Strabismus requires Flamingo and Prickle function to regulate tissue polarity in the *Drosophila* eye. Development 130, 1877–1887.

Shimada, Y., Usui, T., Yanagawa, S., Takeichi, M., Uemura, T. 2001. Asymmetric co-localisation of Flamingo, a seven-pass transmembrane cadherin, and Dishevelled in planar cell polarisation. Curr. Biol. 11, 859–863.

Stern, C., Bridges, C.B. 1926. The mutants of the extreme left end of the second chromosome of *Drosophila melanogaster*. Genetics 11, 503–530.

Struhl, G., Barbash, D.A., Lawrence, P.A. 1997a. Hedgehog acts by distinct gradient and signal relay mechanisms to organise cell type and cell polarity in the *Drosophila* abdomen. Development 124, 2155–2165.

Struhl, G., Barbash, D.A., Lawrence, P.A. 1997b. Hedgehog organises the pattern and polarity of epidermal cells in the *Drosophila* abdomen. Development 124, 2143–2154.

Strutt, D. 2003. Frizzled signaling and cell polarisation in *Drosophila* and vertebrates. Development 130, 4501–4513.

Strutt, D., Johnson, R., Cooper, K., Bray, S. 2002. Asymmetric localisation of Frizzled and the determination of Notch-dependent cell fate in the *Drosophila* eye. Curr. Biol. 12, 813–824.

Strutt, D.I. 2001. Asymmetric localisation of Frizzled and the establishment of cell polarity in the *Drosophila* wing. Mol. Cell 7, 367–375.

Strutt, H., Mundy, J., Hofstra, K., Strutt, D. 2004. Cleavage and secretion is not required for Four-jointed function in *Drosophila* patterning. Development 131, 881–890.

Strutt, H., Strutt, D. 1999. Polarity determination in the *Drosophila* eye. Curr. Opin. Genet. Dev. 9, 442–446.

Strutt, H., Strutt, D. 2002. Nonautonomous planar polarity patterning in *Drosophila*: *disheveled*-independent functions of *frizzled*. Dev. Cell 3, 851–863.

Stumpf, H.F. 1966. Über gefälleabhängige Bildungen des Insektensegmentes. J. Insect Physiol. 12, 601–617.

Taylor, J., Abramova, N., Charlton, J., Adler, P.N. 1998. *Van Gogh*: A new *Drosophila* tissue polarity gene. Genetics 150, 199–210.

Theisen, H., Purcell, J., Bennett, M., Kansagara, D., Syed, A., Marsh, J.L. 1994. *disheveled* is required during *wingless* signaling to establish both cell polarity and cell identity. Development 120, 347–360.

Tokunaga, C., Gerhart, J.C. 1976. The effect of growth and joint formation on bristle pattern in *D. melanogaster*. J. Exp. Zool. 198, 79–96.

Tree, D.R.P., Shulman, J.M., Rousset, R., Scott, M.P., Gubb, D., Axelrod, J.D. 2002. Prickle mediates feedback amplification to generate asymmetric planar cell polarity signaling. Cell 109, 371–381.

Tucker, J.B. 1981. Cytoskeletal coordination and intercellular signaling during metazoan embryogenesis. J. Embryol. exp. Morph. 65, 1–25.

Usui, T., Shima, Y., Shimada, Y., Hirano, S., Burgess, R.W., Schwarz, T.L., Takeichi, M., Uemura, T. 1999. Flamingo, a seven-pass transmembrane cadherin, regulates planar cell polarity under the control of Frizzled. Cell 98, 585–595.

Villano, J.L., Katz, F.N. 1995. *four-jointed* is required for intermediate growth in the proximal-distal axis in *Drosophila*. Development 121, 2767–2777.

Vinson, C.R., Adler, P.N. 1987. Directional non-cell autonomy and the transmission of polarity information by the *frizzled* gene of *Drosophila*. Nature 329, 549–551.

Waddington, C.H. 1943. The development of some "leg genes" in *Drosophila*. J. Genet. 45, 29–43.

Wehrli, M., Tomlinson, A. 1998. Independent regulation of anterior/posterior and equatorial/polar polarity in the *Drosophila* eye; evidence for the involvement of Wnt signaling in the equatorial/polar axis. Development 125, 1421–1432.

Wolff, T., Rubin, G. 1998. *strabismus*, a novel gene that regulates tissue polarity and cell fate decisions in *Drosophila*. Development 125, 1149–1159.

Wright, D.A., Lawrence, P.A. 1981. Regeneration of segment boundaries in *Oncopeltus*: Cell lineage. Dev. Biol. 85, 328–333.

Yang, C.-h., Axelrod, J.D., Simon, M.A. 2002. Regulation of Frizzled by Fat-like cadherins during planar polarity signaling in the *Drosophila* compound eye. Cell 108, 675–688.

Zeidler, M.P., Perrimon, N., Strutt, D.I. 1999a. The *four-jointed* gene is required in the *Drosophila* eye for ommatidial polarity specification. Curr. Biol. 9, 1363–1372.

Zeidler, M.P., Perrimon, N., Strutt, D.I. 1999b. Polarity determination in the *Drosophila* eye: A novel rôle for Unpaired and JAK/STAT signaling. Genes Dev. 13, 1342–1353.

Zeidler, M.P., Perrimon, N., Strutt, D.I. 2000. Multiple rôles for *four-jointed* in planar polarity and limb patterning. Dev. Biol. 228, 181–196.

Zhang, S., Xu, L., Lee, J., Xu, T. 2002. *Drosophila atrophin* homolog functions as a transcriptional corepressor in multiple developmental processes. Cell 108, 45–56.

Zheng, L., Zhang, J., Carthew, R.W. 1995. *frizzled* regulates mirror-symmetric pattern formation in the *Drosophila* eye. Development 121, 3045–3055.

Dorsoventral boundary for organizing growth and planar polarity in the *Drosophila* eye

Amit Singh, Janghoo Lim and Kwang-Wook Choi

Department of Molecular and Cellular Biology, Program of Developmental Biology,
Department of Ophthalmology, Baylor College of Medicine, Houston, Texas 77030

Contents

Advances in Developmental Biology
Volume 14 ISSN 1574-3349
DOI: 10.1016/S1574-3349(04)14004-0

Summary

A fundamental feature of developing tissues and organs is generation of planar polarity of cells in an epithelium with respect to the body axis. The *Drosophila* compound eye shows two-tier dorsoventral (DV) planar polarity. At the individual ommatidium level, the eight photoreceptors in each unit eye form a dorsoventrally asymmetric cluster. At the level of eye field, hundreds of ommatidia in the upper and lower halves of an eye are uniformly polarized dorsally or ventrally, respectively. This results in DV mirror symmetries about the equator. The uniform orientations of photoreceptor clusters over long distance in the eye field provide an excellent model for studying the genetic basis of long-range planar polarity. Ommatidial DV polarity can be detected in third instar eye imaginal disc during the early stage of retinal differentiation. Recent studies have strongly suggested that the foundation for this DV polarity pattern is laid much earlier in undifferentiated eye disc. The eye disc primordium is partitioned into the DV compartments of independent cell lineages. The *Iroquois-Complex* genes specify the dorsal fate whereas their absence in the ventral compartment results in the ventral fate with expression of the *fringe* gene. Genetic evidence suggests that the interaction of dorsal and ventral cells at the DV boundary in early eye disc generates an organizing center for growth and patterning of the eye field. In this chapter, we outline key genetic events involved in early DV patterning and growth of eye disc, and its potential role in organizing long-range signaling for DV planar polarity during later differentiation of the eye.

1. Introduction

Axial patterning is essential for organizing the bodies and organs in animal development. Molecular and genetic basis of anterior–posterior (AP) and DV patterning has been extensively studied in *Drosophila* limb imaginal discs, the larval primordia for adult wings and legs. On the contrary, axial patterning of the eye disc has not been well studied until recently. The aim of this chapter is to present an overview of recent advances in early DV patterning of the *Drosophila* eye disc and its role in organizing planar polarity of differentiating retina.

The adult compound eye consists of approximately 800 ommatidia or unit eyes. Each ommatidium is a honeycomb-like hexagonal facet that contains eight photoreceptor neurons assembled in an asymmetric trapezoidal pattern and the surrounding non-neuronal cell types including pigment and cone cells (Fig. 1; Wolff and Ready, 1993). This stereotypic pattern of cellular organization is repeated in all arrays of ommatidia in the eye. Despite the identical structure of each ommatidium, there is a remarkable difference in the dorsal and ventral halves of the eye: the orientations of asymmetric photoreceptor clusters in each half are polarized in opposite directions, resulting in the formation of a mirror symmetric pattern. There are two main issues in studying the mechanisms of ommatidial planar polarity in the eye. One is to understand how each photoreceptor cluster makes a decision to take the dorsal or ventral

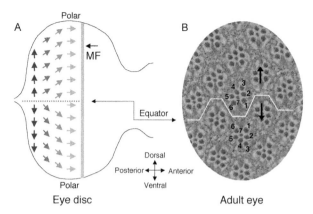

Fig. 1. Dorsoventral mirror symmetry in the eye. (A) Schematic presentation of DV planar polarity of photoreceptor clusters in third instar eye disc. The dorsal and ventral forms of photoreceptor clusters are indicated by arrows pointing to the dorsal and ventral poles, respectively. The morphogenetic furrow (MF) is indicated by a green line. The orientations of the arrows indicate that photoreceptor clusters in the dorsal and ventral halves rotate 90° in opposite directions during maturation of the clusters. (B) Adult eye section in the equator region. Photoreceptors R1 to R7 are indicated by numbers. The R8 cell is not shown in this section as it is located underneath the R7. The dark region of each photoreceptor is the rhabdomere. Note that photoreceptors in each ommatidium (unit eye) are clustered in the asymmetric trapezoidal pattern. Photoreceptor clusters in the dorsal (above the equator line in the middle) and the ventral (below the equator) are mirror-symmetric.

polarity. Recent studies have revealed that asymmetric signal transduction mediated by Frizzled (Fz) receptor for wg/wnt secreted morphogens and Notch (N) signaling plays a key role in determining planar polarity of ommatidia as well as wing hairs (Tomlinson et al., 1997; Boutros et al., 1998; Fanto and Mlodzik, 1999; Tomlinson and Struhl, 1999; Adler, 2002; Strutt et al., 2002; Strutt and Strutt, 2002a; Tree et al., 2002).

The other issue is to find out how the polarity signals are organized to establish long-range planar polarity. The DV mirror symmetry in the eye raises an interesting possibility that the DV planar polarity in the adult eye may result from the DV axial pattern established in the eye disc primordium. In development of the wing and the leg, the imaginal discs are first divided into anterior and posterior compartments of independent cell lineages before they are subdivided into DV compartments (Cohen, 1993). In contrast, there is no significant anterior–posterior lineage restriction boundary in the developing eye disc. However, evidence suggests that major aspects of the DV patterning mechanism are highly conserved in the eye and the wing. An important common finding is that the border between DV compartments is a center for organizing the growth and patterning of the disc (Irvine, 1999).

In the following sections, first we will examine the compartmental nature of DV domains of the eye and genetic basis for the establishment of the DV pattern. We will then address the question of how the early DV pattern organizes growth of the eye disc. Lastly, we will focus on the potential role of early DV pattern in organizing the long-range polarity signal(s).

2. DV compartments and the equator

The equator is the boundary between the dorsal and ventral photoreceptor clusters. Similar mirror symmetric patterns of dorsal and ventral ommatidia in various insect eyes had been described about 100 years ago (Dietrich, 1909). However, developmental mechanisms underlying the DV pattern of these insect eyes have not been well characterized. The use of the *Drosophila* eye as a model has been pivotal in attempts to unravel the genetic basis of DV patterning in the eye.

The adult eye develops from the larval eye imaginal disc, a sac-like epithelial primordium. The eye disc grows without differentiation until early third instar. Retinal differentiation begins from the posterior margin of the eye disc, and the wave of differentiation marked by the morphogenetic furrow proceeds anteriorly resulting in the formation of columns of photoreceptor clusters at regular spacing (Ready et al., 1976). Posterior to the furrow, photoreceptor clusters are generated by a sequence of events including the selection of R8 founder neuron, recruitments of additional photoreceptor precursors in the order of R2/5, R3/4 and R1/6/7 (Tomlinson, 1990; Wolff and Ready, 1993) and 90° rotation (Choi and Benzer, 1994). Preclusters containing five R-cells located a few columns posterior to the furrow already show DV polarity, indicating that the polarity decisions are made during the early differentiation stage (Tomlinson, 1988; Wolff and Ready, 1991). An intriguing question is whether the dorsal and ventral domains of the adult eye are extensions of DV domains observed in early eye discs.

The relationship between the equator and the DV compartmental boundary has been addressed in the pioneering study on *Drosophila* eye development by Ready et al. (1976). By inducing mitotic recombinations between the *white$^+$* (w^+) wild-type and the w^- mutant chromosomes, genetic mosaic eyes were generated that contain clones of w^-/w^- and w^+/w^+ recombinant cells surrounded by w^-/w^+ parental heterozygous cells. The w^+ gene is essential for uptake of red eye pigments and is a useful cell-autonomous marker for photoreceptors and pigment cells (Ready et al., 1976; Lawrence and Green, 1979). The examination of mosaic eyes showed that w^- clones originated from the dorsal side can cross a few cells into the ventral side of the equator and vice versa (Fig. 2). Based on this result, it was concluded that the equator is not determined as the boundary between the dorsal and ventral cell lineages. However, this result does not exclude the possibility that the dorsal and ventral domains of an eye are derived from two independent cell lineages, even though the lineage boundary may not precisely correspond to the equatorial line of DV polarity. This idea was supported by the fact that similar mosaic clones marked by the lack of eye pigmentation due to w^- mutation often show straight clonal border near the DV midline, suggesting that clonal cells are restricted within dorsal and ventral domains without extensive violation of the DV border (Baker, 1967, Campos-Ortega and Waitz, 1978; Lawrence and Green, 1979; Held 2002).

The implication that the eye is derived from DV compartments was further supported by another series of elegant genetic analyses of a large number of mosaic clones in the adult eye and head (Baker, 1978). In an attempt to see whether the eye and head are also subdivided into different domains by sequential compartmentalization

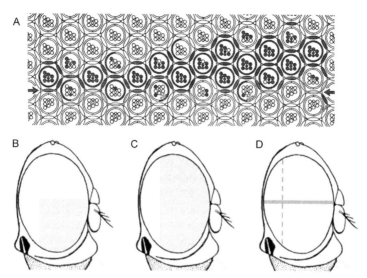

Fig. 2. Clonal restrictions in the eye. Schematic of mosaic clones in the adult eye. (A) Homozygous w^- clone (shown in black) in the equator region. Mosaic clones were generated by X-ray-induced mitotic recombination during first instar larval stage. Ommatidia at the clone border consist of both w^+ and w^- photoreceptors, indicating that the eight photoreceptors in each ommatidium are not clonally related. Note that most w^- ommatidia are located in the dorsal side but a few w^- cells marked by red asterisks can be found in the ventral ommatidia below the equator (arrows). Adapted from Ready et al. (1976). (B–D) Representative large homozygous M^+/M^+ clones (the pink area). (B) This clone shows straight clone boundary at the dorsal ventral midline. (C) This clone shows straight AP clone boundary occasionally detected at about one-third from the posterior end of the eye. There is a weak but potential compartmental restriction whose significance is unknown. (D) It shows two potential cell lineage restriction lines. The strongest DV lineage boundary and a weak potential AP boundary are indicated by a blue line and a purple dashed line, respectively. Adapted from Baker (1978). (See Color Insert.)

as in the wing, mosaic analysis was carried out using *Minute* (*M*) mutation that dominantly reduces growth rate of M^-/M^+ cells. Clones of M^+ cells generated by mitotic recombinations in M^-/M^+ heterozygous animals have a growth advantage due to a higher cell division rate than that of surrounding M^-/M^+ cells (Morata and Ripoll, 1975, Garcia-Bellido et al., 1976), thus determining whether there are compartmental boundaries into which M^+ clonal cells cannot invade.

Consistent with the results from w^- mosaic studies, nearly all (96%) large M^+ clones generated during the early larval stage (24–41 h after egg laying, AEL) showed sharp clone borders near the DV midline indicating that clones are restricted to either the dorsal or ventral domain (Baker, 1978). DV lineage restriction shown in the adult eye was also confirmed in the developing eye disc by identifying M^+ clones by the lack of a molecular marker, CD2. Compartments of different cell lineages do not intermingle due to differences in cell affinity (Garcia-Bellido et al., 1973; Crick and Lawrence, 1975; Irvine, 1999). Conversely, two groups of cells with the same cell identity tend to intermingle with each other. In eye discs containing M^+ clones induced during 48–72 h AEL showed straight lines of clone boundaries along the

DV midline whereas clones located within the dorsal or ventral domain had wiggly borders (Dominguez and de Celis, 1998). Furthermore, immunocytochemical staining of photoreceptor nuclei indicated that the straight clone borders approximately corresponds to the equator. These results support that there is a straight line of clonal restriction between dorsal and ventral ommatidia.

Although clonal analysis strongly suggests that there is a compartment boundary between the dorsal and ventral domains, observations of w^- clones trespassing the equator (Ready et al., 1976) indicate that clone borders do not precisely correspond to the equator. Consistent with this, analysis of M^+/M^+ clones in eye discs also indicates that ommatidia located at the DV restriction line consist of both dorsal and ventral cells (Dominguez and de Celis, 1998). However, some degree of straddling of a clone across the compartment boundary has also been found in other imaginal discs. For example, the bristles of a tarsal row 1 in the leg straddles the AP boundary and clonal cells can be embraced by either A or P compartment depending on the way that sensory organ precursors (SOPs) are stochastically selected (Lawrence et al., 1979; Held, 2002). Therefore, it is possible that DV compartments formed in the eye disc during the early larval stage generate an approximate DV boundary, and the cells straddling the DV boundary may be recruited to either the dorsal or ventral ommatidia across the equator by the way that photoreceptors are selected through local cell–cell interactions during differentiation.

3. Dorsal selector genes

3.1. Identification of Iroquois-Complex (Iro-C)

In wing imaginal disc, the identity of compartments is specified by selector genes such as *engrailed* (*en*) and *apterous* (*ap*) that are expressed in the posterior and dorsal compartments, respectively (Brower, 1986; Cohen et al., 1992; Blair, 1995; Hidalgo, 1998). If the eye disc consists of distinct DV compartments, it is expected that there are genes specifically expressed in the dorsal or ventral compartment. Enhancer trap technique using P-element containing *mini-white* (*w*) and *lacZ* reporter gene (*P-lacW*) has been instrumental in identifying genes expressed in specific spatial patterns (Bier et al., 1989). Enhancer trap lines that show a dorsal-specific w^+ expression pattern led to the identification of dorsal selector genes in the eye (Sun et al., 1995; Choi et al., 1996; McNeill et al., 1997; Kehl et al., 1998; Netter et al., 1998; Table 1). Interestingly, most dorsal-specific *P-lacW* enhancer traps isolated from independent genetic screens map to a single chromosomal region, identifying this region as a key site for dorsal-specific genes in the eye.

In this chromosomal region, a cluster of three conserved homeobox-containing genes, *araucan* (*ara*), *caupolican* (*caup*), and *mirror* (*mirr*) were identified (Gomez-Skarmeta et al., 1996; McNeill et al., 1997; Grillenzoni et al., 1998; Kehl et al., 1998). Since mutants in these genes lack lateral thoracic bristles, these genes were named the *Iroquois-Complex* (*Iro-C*) based on the hairstyle of the Indian tribe (Gomez-Skarmeta et al., 1996; Grillenzoni et al., 1998). These three genes of the *Iro-C* are clustered within an approximately 140 kb region (Netter et al., 1998). Vertebrate homologs of *Iro-C* genes also exist in genomic clusters of three similar genes and play roles in defining territories in various tissues (Bosse et al., 1997; Cavodeassi et al., 2001). All three *Drosophila Iro-C* genes are expressed in the dorsal domain of the eye disc, playing crucial roles in the specification of the dorsal fate as described below.

3.2. Dorsal selector function of Iro-C in head-eye compartments

Mutations in each member of *Iro-C* complex cause lethality, indicating that each of *Iro-C* genes plays an essential role during early development. Despite similar expression patterns in imaginal discs, significant differences were found in the expression pattern of *Iro-C* genes in other tissues. *mirr* is strongly and dynamically expressed in the CNS (Netter et al., 1998; Urbach and Technau, 2003) while *ara* and *caup* are preferentially expressed in mesodermal tissues in the embryos (Netter et al., 1998). *mirr* is essential for follicle cell patterning in oogenesis by controlling *pipe* (*pip*), a gene involved in the establishment of the DV axis of the embryo (Jordan et al., 2000; Zhao et al., 2000). However, in the eye disc, all three *Iro-C* members are expressed in the dorsal domain in a nearly identical pattern, suggesting that their function might be redundant in the eye. Analysis of the *mirr* mutant reveals weak but significant polarity defects in the eye (McNeill et al., 1997). Loss of *mirr* causes nonautonomous DV polarity reversals, that is, the DV polarity in *mirr*$^+$ ommatidia adjacent to a *mirr*$^-$ clone was reverted. This phenotype was only seen when the clones were generated in the dorsal side of the eye, which is consistent with the restricted expression of *mirr* in the dorsal domain. The nonautonomous polarity reversals along the clone border suggests that the boundary between *mirr*$^+$ and *mirr*$^-$ cells is an important determinant for the generation of the equator and that *mirr* may act as a selector for the dorsal fate in the eye disc. Clones of *mirr*$^-$ mutant cells show round and smooth clone borders, indicating that *mirr*$^-$ cells avoid mixing with the neighboring *mirr*$^+$ cells. In contrast, clones of *mirr*$^+$ wild-type control cells surrounded by *mirr*$^+$ cells show wiggly clone borders, suggesting that a *mirr*$^+$ clone can interact normally with neighboring cells (Yang et al., 1999). This analysis of clone shapes suggests that *mirr* is involved in the specification of the dorsal identity and in the establishment of the dorsal compartment.

An important function of compartment boundaries in imaginal discs is to organize a signaling event necessary for patterning and growth of discs. Although loss-of-function *mirr* mutant clones were sufficient to induce polarity pattern, the effects were limited to relatively short-range and there was no indication that *mirr*$^+$/*mirr* border induces growth of the disc. As discussed above, two other dorsal genes of *Iro-C*, *ara* and *caup*, may partially compensate the loss of *mirr* function in the eye as they are expressed in the same dorsal domain. To resolve this issue of functional redundancy, a deletion mutation *iro*^{DMF3}, which affects all three *Iro-C* genes by the deletion of *ara* and *caup* as well as a 5′-region of *mirr* (Gomez-Skarmeta et al., 1996; Diez del Corral et al., 1999), was used for clonal analysis. In contrast to *mirr*$^-$ clones, *Iro-C* clones in the dorsal side of the eye induced either enlargement or ectopic eye field (Cavodeassi

Table 1
Genes involved in early DV patterning and domain-specific growth

	Gene product	Function	References
Dorsal			
araucan (*ara*)	Homeodomain	Dorsal selector	Gomez-Skarmeta (1996), Cavodeassi et al. (1999, 2000), Pichaud and Casares (2000)
caupolican (*caup*)	Homeodomain	Dorsal selector	Gomez-Skarmeta (1996), Cavodeassi et al. (1999, 2000), Pichaud and Casares (2000)
mirror (*mirr*)	Homeodomain	Dorsal selector	McNeill et al. (1997), Heberlein et al. (1998), Kehl et al. (1998), Yang et al. (1999)
pannier (*pnr*)	GATA family	Dorsal selector	Ramain et al. (1993), Maurel-Zaffran and Treisman (2000)
wingless (*wg*)	Secreted	Control of *mirror*	Ma and Moses (1995), Treisman and Rubin (1995), Heberlein et al. (1998), Wehrli and Tomlinson (1998)
Delta (*dl*)	Transmembrane Notch ligand	Dorsal N ligand	Alton et al., (1989), Sun et al. (1997), Cho and Choi (1998), Dominguez and de Celis (1998), Pappayannopoulos et al. (1998)
Ventral			
fringe (*fng*)	Glycosyltransferase	DV boundary	Irvine and Wieschaus (1994), Cho and Choi (1998), Dominguez and de Celis (1998), Pappayannopoulos et al. (1998)

Serrate (ser)	Transmembrane Notch ligand	Ventral N ligand, ventral growth	Speicher et al. (1994), Cho and Choi (1998), Domínguez and de Celis (1998), Pappayannopoulos et al., (1998), Cho et al. (2000)
decapentaplegic (dpp)	TGF-β	Ventral growth	Masucci and Miltenberger (1990), Wiersdorff et al. (1996), Chanut and Heberlein (1997)
Equatorial			
WRI22	Unknown	Unknown	Heberlein et al. (1998)
four-jointed (fj)	Type-II membrane glycoprotein	Eq-pole gradient	Villano and Katz (1995), Brodsky and Steller (1996)
equator1 (eq1)	Unknown	Unknown	Sun et al. (1995)
equator2 (eq2)	Unknown	Unknown	Sun et al. (1995)
Asymmetric function			
Lobe (L)	Novel protein	Ventral growth	Chern and Choi (2002)
teashirt (tsh)	Homeodomain	Dorsal growth, ventral suppression	Fasano et al. (1991), Pan and Rubin (1998), Bessa et al. (2002), Singh et al. (2002)
homothorax (hth)	Homeodomain	Ventral suppression	Rieckhof et al. (1997), Pai et al. (1998), Jaw et al. (2000), Pichaud and Casares (2000)

et al., 1999, 2000; Pichaud and Casares, 2000). These clones were also able to induce repolarization of ommatidial polarity over a long range from the clone border (Fig. 3). This result suggests that ectopic DV boundary between *Iro-C*$^+$ and *Iro-C*$^-$ cells generated by *Iro-C* mutant clone in the dorsal domain of eye is sufficient to organize polarity pattern and growth, fulfilling the condition of compartment boundary. Furthermore, ectopic *Iro-C* border generated by misexpression of any *Iro-C* gene in the ventral region of the eye is sufficient to reorganize DV polarity and to promote the formation of ectopic eye fields, albeit at a low frequency (Cavodeassi et al., 2000). Taken together, these results strongly support that the *Iro-C* complex as a whole is essential for organizing growth and DV polarity pattern of the eye, although the three *Iro-C* genes are partially redundant.

A composite eye-antenna disc gives rise to the eye and antenna. This disc also contains subregions that give rise to the eye and head structures including ptilinum, frons, and maxillary palpus. These subregions of the eye-antenna disc originate from

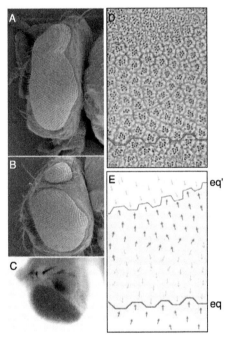

Fig. 3. *Iro-C* function as a dorsal selector. Consequences of removing *Iro-C* activity in eye development. In all panels, anterior is to the left and dorsal is up. (A–C) Adult eyes harboring dorsal *Iro-C* clones. Mutant tissue is genetically labeled by the *white* mutation, appearing as a pigmentless tissue against the red-pigmented wild-type tissue. (D) Section through an eye carrying a dorsal (upper *white* tissue) and a ventral (lower *white* tissue) clone. (E) Schematic representation of the ommatidial polarity of the eye in (D). Dorsal (blue) and ventral (red) ommatidial chirality is represented by arrows. The equator (eq) is outlined by a thick blue line in D and E. The border of the dorsal *Iro-C* clone defines an ectopic equator (eq', in E). The ventrally located clone does not show phenotypic effect. Adapted from Cavodeassi et al. (2000). (See Color Insert.)

six embryonic segments (Struhl, 1981; Jurgens and Hartenstein, 1993). Clonal removal of *Iro-C* in mosaic animals can transform the dorsal head capsule into ventral head in cell-autonomous manner (Cavodeassi et al., 2000). Therefore, *Iro-C* genes are required for specification of dorsal head structures as well as the dorsal eye. Induction of mitotic recombination at different times of development results in transformation of different structures. *Iro-C* clones induced during late larval stages (48–72 or 72–96 h AEL) tend to generate only the ventral head parts without ectopic eyes or enlarged eyes. Conversely, most common transformations in early induced clones (24–48 h AEL) are ectopic eyes or dorsal enlargement of the eye (Cavodeassi et al., 1999, 2000; Pichaud and Casares, 2000). This suggests that DV patterning of the eye occurs in earlier larval stages than the head patterning.

4. Dorsoventral boundary as an organizing center

It has been shown in the wing that LIM-homeodomain protein Apterous (Ap) acts as a dorsal selector and induces expression of Fringe (Fng) and Serrate (Ser) in the dorsal compartment (Cohen et al., 1992; Diaz-Benzomea and Cohen, 1995; de Celis et al., 1996; Bachmann and Knust, 1998). *ser* is a ligand for Notch (N) in the dorsal cells whereas Delta (Dl) is the N-ligand in the ventral cells. Fng is critical for restricted N activation at the DV border (Irvine and Wieschaus, 1994; Kim et al., 1995; Doherty et al., 1996; Fleming et al., 1997; Klein and Arias, 1998; Wu and Rao., 1999; Lawrence et al., 2000). *fng* directly binds N to promote N-Dl interaction. Fng acts as a glucosyltransferase that elongates O-linked fucose residues to the EGF domains of N and modulates N signaling (Haltiwanger, 2002; Okajima and Irvine, 2002). In contrast to the function of Fng in promoting N-Dl interaction, Fng bound to N protein inhibits Ser-N interaction (Ju et al., 2000; Moloney et al., 2000). This function of Fng allows Dl to activate the N receptor in the dorsal cells only at the DV border. Similarly, Ser can activate the N receptor in the ventral cells only at the DV border since Ser-N interaction is prevented by Fng in the ventral cells away from the DV border (Fig. 4).

Except for the dorsal wing selector Ap, other components involved in N signaling in the wing such as Dl, Ser, and Fng, are expressed in the domain-specific pattern and are required for eye disc patterning. It is noteworthy that Fng is ventral-specific in the eye but dorsal-specific in the wing. Similarly, Dl and Ser are preferentially expressed in the dorsal and ventral domains of the eye disc, respectively. Therefore, the DV axis in the wing is inverted in the eye. It has been proposed that the DV axis in the eye is inverted by 180° rotation of the eye primodium during embryogenesis (Struhl, 1981). This rotation about the left–right axis of the embryo inverts anterior–posterior as well as DV axis of the eye-antenna disc. Despite the inversion of DV axis, the function of Fng seems to be equivalent in both eye and wing discs. Consistent with the role of Fng in the wing, N is activated along the DV midline as indicated by expression of E(Spl)mβ, a transcriptional target of N signaling (Dominguez and de Celis, 1998). The expression pattern of these genes in the developing eye disc changes dynamically, resulting in striking differences before and after the initiation of retinal

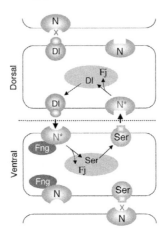

Fig. 4. A model showing Fringe-dependent Notch signaling at the DV boundary in eye Dl and Ser are preferentially expressed in the dorsal and ventral cells, respectively. Fng is expressed in the ventral cells. The binding of Fng to N enhances or reduces its affinity to Dl and Ser, respectively. In the absence of Fng, Dl-N interaction is promoted when N is associated with fng. *dl* cannot activate N in the dorsal cells due to the absence of Fng. In the ventral cells, Ser cannot activate N because N is associated with Fng. Therefore, Dl and Ser can activate N (marked by 'N*') only at the DV boundary. Activated N also induces Fj which might be secreted to form an equator-to-pole gradient for planar polarity signaling. Note that the DV axis in the eye is the inverse to that of the wing. Adapted from Ju et al. (2000).

differentiation. In the case of Fng, it is initially expressed in the ventral domain, but as the eye disc develops further, the ventral specificity is no longer maintained. As eye disc undergoes retinal differentiation, Fng expression is preferentially localized anterior to the furrow (Fig. 5).

The essential role of *fng* in DV patterning was demonstrated by analysis of *fng* mutant clones. When *fng* clones are induced in the ventral but not in the dorsal domain of the eye, DV polarity is reorganized near the ectopic border between *fng*[+] and *fng*[−] cells resulting in polarity reversals over a long distance of several ommatidia (Cho and Choi, 1998; Dominguez and deCelis, 1998; Papayannopoulos et al., 1998). Polarity reversals by *fng* mutant clones were nonautonomous because most were not restricted within the clones but rather found in the wild-type region adjacent to the clone border, resulting in the generation of ectopic equators (Fig. 5). This suggests that localized activation of N at the *fng*[+]/*fng*[−] DV boundary is important for organizing the pattern of planar polarity. Polarity reversals induced by *fng* mutant clones are found preferentially in the equatorial side of a clone, and the effect of polarity reversal is stronger when the *fng*[−] clones are located farther away from the equator. This position-dependent effect suggests that a polarity signal may be induced at the DV boundary in the equator-to-pole gradient and an ectopic DV boundary formed by a *fng* mutant clone may induce an ectopic polarity gradient. In this case, *fng* mutant clones far away from the equator may have less influences from the endogenous signal and therefore an ectopic signal is likely to have stronger effects of repolarization. (see also Section 7).

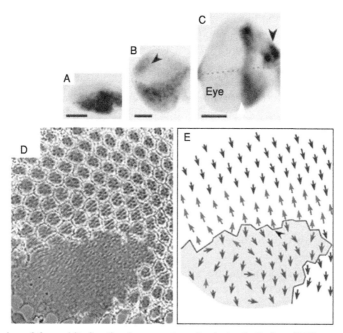

Fig. 5. Expression of *fng* and its function in generation of polarity A–C, *fng* mRNA expression. Dotted lines indicate DV midline (A–C). (A) First instar disc, (B) late second instar disc, (C) Late third instar disc. Arrowheads indicate weak *fng* expression in dorsal domain (B) and expression in the ocelli region (C), respectively. Equatorial region and morphogenetic furrow are marked with black and white arrows, respectively (C). Scale bars: a, 5 μm; b, 15 μm; c, 45 μm. (D) Section of *fng* mosaic eye showing polarity reversals by ventral *fng⁻* clones (D). (E) Schematic presentaion of (D). *fng⁻* clones are colored green. Dorsal and ventral trapezoids are indicated by red and blue arrows, respectively. *fng⁻* clone near the ventral margin shows polarity reversals in 3–4 rows of ommatidia outside the clone border. An ommatidium with abnormal rotation but normal chirality is marked with a black arrowhead. Anterior is to the right. Adapted from Cho and Choi (1998). (See Color Insert.)

Activated N signaling induced by *fng⁺*/*fng⁻* border promotes disc growth (de Celis et al., 1996; Go et al., 1998; Baonza and Garcia-Bellido, 2000). Misexpression of constitutively active intracellular domain of N (N^{intra}) causes overgrowth of the eye as well as the wing disc (Dominguez and de Celis, 1998; Go et al., 1998; Papayanno-poulos et al., 1998; Baonza and Garcia-Bellido, 2000; Chern and Choi, 2002). Con-versely, loss of N leads to a failure of eye disc growth. Overexpression of *fng* in both the dorsal and ventral eye using the Gal4-UAS system (Brand and Perrimon, 1993) leads to strong reduction or complete loss of the eye, suggesting that loss of border between *fng⁺* and *fng⁻* cells causes disruption of disc growth and/or differentiation of the eye disc. Hence, results from loss-of-function and overexpression of *fng* suggest that N signaling induced by *fng⁺* and *fng⁻* cell boundary is important for organizing planar polarity and growth.

The critical time for *fng* function in DV patterning was estimated by determining when the eye fails to develop by overexpression of *fng* at different times of development.

Since the Gal4 expression level is elevated as the culture temperature increases (Brand and Perrimon, 1993), *fng* was conditionally overexpressed by modulating the Gal4 level by shifting the temperature to 29 °C. Results from temperature shift experiments suggest that the critical time for *fng* function is between the late first instar and early second instar stage (Cho and Choi, 1998). Several genes including *fng*, *Ser*, and *Dl* are preferentially expressed transiently in dorsal or ventral domain of eye discs during first and second instar stages (Cho and Choi, 1998; Dominguez and de Celis, 1998; Papayannopoulos et al., 1998). Together with previous mosaic analyses discussed earlier, these studies lead to a similar conclusion that the critical time for establishing DV pattern is approximately between the first and early second instar.

5. Regulation of domain-specific genes

5.1. Control of Iro-C by wingless and pannier

Iro-C gene expression in the dorsal compartment is crucial for growth and patterning of planar polarity in the eye. How is *Iro-C* expression regulated? One of the candidate genes that might control *Iro-C* expression is *wingless* (*wg*) because it is expressed in the dorsal polar region in early disc (Cavodeassi et al., 2000; Cho et al., 2000). To examine whether *wg* is involved in dorsal-specific expression of *Iro-C*, a temperature-sensitive *wg* mutant allele wg^{ILII4} was used to inhibit *wg* function at different times during development. The position of DV midline in the eye disc can be marked by the Bolwig's nerves that run along the DV midline of eye disc to the optic stalk or molecular markers such as *WR122-lacZ* enhancer trap which are expressed broadly in the equator region posterior to the furrow (Heberlein et al., 1998). When the mutant larvae were transferred from a permissive temperature (17 °C) to a restrictive temperature (29 °C) during first instar stage or 48 h prior to dissection at late third instar stage, *WR122-lacZ* was ectopically expressed in the dorsal margin region where ectopic retinal differentiation occurs (Heberlein et al., 1998). Furthermore, misexpression of *wg* was able to repress *WR122-lacZ* expression. This suggests that the expression of *wg* in the DV marginal regions is involved in the restriction of *WR122-lacZ* expression to the equatorial region.

Misexpression of *wg* driven by *dpp-Gal4* results in expansion of *mirr-lacZ* dorsal expression across the endogenous DV midline, whereas *mirr-lacZ* expression is reduced in wg^{ILII4} discs after incubation at the restrictive temperature (Heberlein et al., 1998). These results suggest that *wg* is necessary and sufficient for positioning of the equator by inducing *mirr* expression in the dorsal domain and further restricting the expression of *WR122-lacZ* to the DV midline region. It is important to note that there are significant differences in the effects of *wg* overexpression or wg^{ILII4} mutation depending on the timing of *wg* manipulation. The most striking effects of *wg* were seen when *wg* was reduced during the early larval stage. This suggests that *wg* -dependent positioning of DV midline is established in early discs. Further analysis of *wg* function using mutant clones of *wg* pathway genes indicated that *wg* signaling is essential for *mirr* expression in the dorsal but not in the ventral domain, suggesting that *mirr* is under repression by unknown factors (Cavodeassi et al., 1999; Lee and Treisman, 2001).

It has been shown that *pannier* (*pnr*), a GATA family zinc-finger transcription factor (Ramain et al., 1993), functions as a dorsal selector for patterning of the adult thorax (Calleja et al., 1996; Heitzler et al., 1996; Calleja et al., 2000). Therefore, it was possible that *pnr* might also play a role in the DV patterning of the eye. *pnr* is expressed in the dorsal region of eye/head primordia in the embryo as well as in the dorsal margin of the larval eye disc. (Heitzler et al., 1996; Maurel-Zaffran and Treisman, 2000). *pnr*⁻ mutant clones in the dorsal domain induce enlargement of the dorsal eye or ectopic eye (Maurel-Zaffran and Treisman, 2000). Sections of adult eyes with *pnr* mutant clones show nonautonomous DV polarity reversals and ectopic equators. These results further confirm that the *pnr*⁺/*pnr*⁻ boundary, but not the loss of *pnr* *per se,* is important for causing dorsal enlargement or ectopic eye formation. These phenotypes are remarkably similar to those of *Iro-C*⁻ clones, and suggest that *pnr* acts as a dorsal selector in the eye. On the contrary, when *pnr* mutant clones are generated in M^-/M^+ background, *pnr*⁻ M^+ clonal cells have a growth advantage over surrounding *pnr*⁺ M^-/M^+ cells, resulting in the eye disc in which most or all cells are homozygous *pnr*⁻. In such cases, *pnr*⁻ mutant clones cause a loss of almost all or the entire eye (Maurel-Zaffran and Treisman, 2000). Similarly, eye development is strongly suppressed when *pnr* activity is inhibited by misexpression of *U-shaped* (*Ush*) (Fossett et al., 2001), a *Drosophila* homolog of FOG (the Friend of GATA) family zinc-finger proteins that bind *pnr* to inhibit its transcription activity (Cubadda et al., 1997; Haenlin et al., 1997; Tsang et al., 1997). These results also support that the generation of a boundary between *pnr*⁺ and *pnr*⁻ cells is critical for organizing eye development and that loss of such boundary disrupts development.

It is interesting to note that *pnr* expression is restricted to the dorsal margin throughout eye disc development and never expand to the dorsal half. Therefore, the border between *pnr*⁺ and *pnr*⁻ cells cannot be the DV boundary of the eye. Genetic tests for epistatic relationships between the dorsal genes indicate that the dorsal expression of *wg* and *Iro-C* genes are controlled by *pnr* (Maurel-Zaffran and Treisman, 2000; Lee and Treisman, 2001). Since *wg* is important for defining the dorsal domain of *Iro-C*, the primary function of *pnr* might be the induction of *wg* expression in the cells of the dorsal margin region, although it is unknown whether *wg* is a direct target of *pnr*. Since *wg* is diffusible, the distribution of *wg* in a broader dorsal region may induce the expression of *Iro-C* in the dorsal half and establish the DV border.

5.2. Function of peripodial membrane in DV patterning and growth

Wg and the *Drosophila* homolog of TGF-β. Decapentaplegic (Dpp), are important morphogens that function antagonistically to each other in controlling furrow initiation and progression during initial stage of eye morphogenesis (Ma and Moses, 1995; Treisman and Rubin, 1995; Cadigan et al., 2002). Since *Iro-C* is expressed in the dorsal domain of very early stage eye disc, *wg* is expected to be expressed very early to regulate *Iro-C* expression. Indeed, *wg*-expressing cells marked by *wg-lacZ* are detected in the dorsal part of first instar eye discs (Cho and Choi, 1998). Interestingly, *dpp-lacZ* cells are located in the ventral part in the complementary pattern. The early expression of *wg* and *dpp* is important for DV patterning since overexpression of *wg* or loss of *dpp* leads to disruption of the DV pattern of Dl and Ser that are preferentially expressed in the dorsal and ventral domain, respectively (Cho et al., 2000). It is important to note that imaginal discs are sac-like structures consisting of two apposing epithelial cell layers, a peripodial layer of squamous cells and a disc proper layer of columnar cells. Intriguingly, most cells expressing *wg* and *dpp* are found in the peripodial membrane rather than in the eye disc proper where retinal differentiation takes place. Although the peripodial membrane plays a role in the eversion of imaginal discs into corresponding adult structures (Milner et al., 1983; Fristrom and Fristrom, 1993; Agnes et al., 1999), it does not contribute to the structure of appendages. However, recent studies suggest that peripodial membrane and disc proper cross-talk across the lumen during development, perhaps for coordinating growth of both layers of the disc epithelia (Cho et al., 2000; Gibson and Schubiger, 2000).

Genetic evidence for peripodial signaling was obtained by analyzing peripodial membrane- or disc proper-specific clones. *flp*-out clones of *hh*-expressing cells in the peripodial membrane induced ser across the disc lumen but no such effect was seen when the clones were induced in the disc proper. Clones of *hh* LOF mutation in the peripodial membrane cause loss of underlying tissue in the disc proper. This effect is also due to loss of the *hh* signal originating from the peripodial membrane since such clones in the disc proper show no obvious defects (Cho et al., 2002). In addition to the peripodial-to-disc proper signaling, signaling in the opposite direction also occurs in both wing and eye discs (Gibson et al., 2002). These studies indicate that translumenal signaling of *hh* and *dpp* is important for growth of the disc.

It is also noteworthy that *pnr*, a key upstream regulator, or *Iro-C*, is specifically expressed in the dorsal margin of peripodial cells (Fig. 6). In contrast, Iro-C proteins are expressed not only in the dorsal peripodial margin but also in the disc proper in

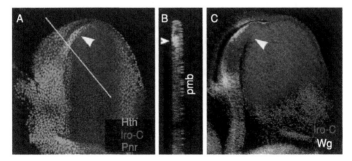

Fig. 6. Peripodial expression of *pnr*. A third instar eye-antennal imaginal disc shows expression of GFP reporter of *pnr* expression driven by *pnr-Gal4/UAS-GFP*. It also shows the expression of *hth* and rF209, a β-galactosidase reporter for *Iro-C*. *pnr* is specifically expressed in the dorsal peripodial cells. *Iro-C-lacZ* is expressed in the dorsal half of the eye disc. It is also detected in the peripodial membrane. (B) Cross section of the disc in (A) at the angle indicated as a line. It shows that *pnr* expression is restricted in the peripodial membrane. *Iro-C-lacZ* shows little overlap with *pnr*. *hth* is also expressed in the peripodial membrane but does not overlap with *pnr*. Adapted from Pichaud and Casares (2000). (See Color Insert.)

the non-polar region. The restricted expression of *pnr* in the dorsal peripodial membrane raises a possibility that Iro-C expression in the disc proper may be mediated by secretion of Wg in the dorsal region. Since Wg is expressed in the peripodial cells in the dorsal margin, *pnr* may be involved in induction of Wg which in turn induces Iro-C expression by translumenal signaling from the peripodial membrane to the disc proper. Peripodial cells have cytoplasmic processes that extend to the disc proper across the lumen (Cho et al., 2000; Gibson and Schubiger, 2000). It is an interesting possibility that Wg itself, or other molecules induced by *pnr*, might be transported to the dorsal disc proper cells to regulate a secondary signaling event to turn on Iro-C expression.

6. Domain-specific growth of eye

6.1. Function of Homothorax and Teashirt

Unlike domain-specific genes discussed above, a new class of genes have been identified that are expressed in both dorsal and ventral domains but display domain-specific functions. An example of this gene class is *homothorax* (*hth*) that encodes a homeodomain protein, an essential component for nuclear localization of Extradenticle (Exd) (Rieckhof et al., 1997; Jaw et al., 2000). Both *hth* and *exd* are negative regulators of eye development since loss of function mutant clones induces ectopic retinal differentiation (Pai et al., 1998). Interestingly, loss of function clones of *hth* and *exd* induce similar ectopic eyes in the ventral side of the eye, whereas dorsal clones do not (Gonzalez-Crespo and Morata, 1995; Pai et al., 1998). *hth* is expressed ubiquitously in early eye disc during second instar stage but becomes more restricted in the DV margin anterior to the furrow in differentiating third instar eye disc. *wg* is also expressed in this anterior DV margin region. Analysis of *hth* mutant clones and *hth* misexpression indicates that *hth* is required for maintenance of ventral *wg* expression (Pichaud and Casares, 2000). Since *wg* expression anterior to the furrow antagonizes furrow initiation (Ma and Moses, 1995; Treisman and Rubin, 1995), ectopic eye formation resulted from loss of *hth* is likely to be due to a reduced *wg* level in the ventral region of the eye disc.

Another homeotic gene, *teashirt* (*tsh*), is known to be involved in conferring the proximal identity in the leg and wing due to its ability to induce *hth* expression (Erkner et al., 1999; Wu and Cohen, 2002). In the eye, *tsh* was shown to act downstream to *eyeless* (*ey*) and upstream to *eyes absent* (*eya*), *sine oculis* (*so*), and *dachshund* (*dac*), all of which are important for retinal fate determination and are sufficient to induce ectopic eyes upon ectopic expression (Kumar and Moses, 2001a). Therefore, *tsh* may be part of the genetic network that functions to specify eye identity (Pan and Rubin, 1998). Like *hth* in the eye, *tsh* is also expressed in both dorsal and ventral domains of the eye disc, but its loss of function results in dorsoventrally asymmetric phenotypes: dorsal reduction vs. ventral overgrowth (Singh et al., 2002). Gain-of-function clones of *tsh* generated by *flp*-out technique (Ito et al., 1997) in the dorsal region of the eye result in the dorsal enlargements whereas the clones on the ventral eye margin show strong suppression of the eye. These domain-specific phenotypes differ from those of *hth* described earlier as neither loss nor gain of *hth* function results in the dorsal phenotypes as seen in the case of *tsh* despite similar ventral phenotypes (Fig. 7). Further analysis has revealed that the function of *tsh* in ventral suppression of eye development was due to its ability to induce *hth*, a *wg*-dependent negative regulator of eye development.

Since *hth* and *tsh* are expressed in both dorsal and ventral domains of eye disc, their domain-dependent functions might require some additional spatial cues for generation of DV asymmetry. One possible mechanism for the asymmetric functions of *hth* and *tsh* is that these genes interact with DV pattern genes (Fig. 7). Indeed, the function of *tsh* in the promotion of dorsal eye development is dependent on the presence of *Iro-C*, whereas its opposite function of suppressing ventral eye development requires Ser, an N ligand preferentially expressed in the ventral domain of early eye disc (Singh et al., 2004). It will be interesting to see whether Tsh directly interacts with Iro-C or unknown targets of Iro-C transcription factors to promote dorsal eye development. Alternatively, Hth and Tsh may be differentially modified to become active or inactive in the domain-specific manner.

6.2. Function of Decapentaplegic in ventral growth

In addition to Hth and Tsh transcription factors that function in a dorsoventrally asymmetric manner, there is other striking evidence suggesting that growth of eye disc is asymmetrically or independently regulated in a DV compartment-specific manner. For instance, dpp^{blk}, an eye-specific *dpp* allele that has a deletion in the 3' eye-specific enhancer region, shows preferential loss of ventral eye tissue (Masucci

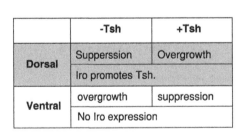
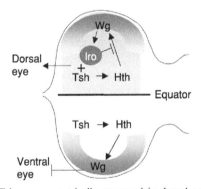

Fig. 7. Asymmetric function of Hth and Tsh. Hth and Tsh are symmetrically expressed in dorsal and ventral domains but show distinct functions in each domain. In the ventral domain, Tsh induces Hth expression which is necessary for maintain the Wg level. In the absence of either Tsh or Hth, the ventral eye region overgrows due to reduction of Wg. Conversely, overexpression of either Tsh or Hth leads to suppression of ventral region. In contrast to the similar function of Tsh and Hth in the ventral domain, Tsh and Hth have different functions in the dorsal domain. Hth function in the dorsal region is inhibited in the presence of a dorsal component such as Iro-C. However, Tsh and Iro-C together promote dorsal eye growth. Summary of the consequences of loss or overexpression of Tsh in dorsal and ventral domains is shown in the left panel.

and Miltenberger, 1990). *dpp* is involved in diverse aspects of disc growth and patterning (Neumann and Cohen, 1997; Raftery and Sutherland, 1999). In differentiating eye disc, *dpp* is expressed in the furrow under the control of *hedgehog* (*hh*) and is required for furrow progression. In *dpp^{blk}* mutant, the eye disc appears to grow to a nearly normal size prior to the onset of differentiation and shows no significant changes in the expression of dorsal and ventral markers, indication that growth and patterning in early larval stages are relatively normal. However, the expression of *dpp-lacZ* reporter is reduced in the ventral margin, and the furrow progression is strongly retarded or absent in the ventral side, resulting in a restriction of ommatidial development to the dorsal side (Wiersdorff et al., 1996; Chanut and Heberlein, 1997). This defect can be rescued by elevating the *dpp* level by inducing *ptc⁻* mutant clones in the ventral eye margin region. As Ptc is the Hh receptor that negatively regulates Hh signaling, loss of Ptc induces ectopic Hh and therefore Dpp expression. Therefore, the primary defect in *dpp^{blk}* appears to be the failure of furrow initiation in the ventral side of eye disc (Chanut and Heberlein, 1997).

Analysis of *dpp-lacZ* reporter expression throughout larval development has revealed dynamic changes in the pattern of expression (Cho et al., 2000). Interestingly, *dpp-lacZ* expression during first instar is ventral-specific, and becomes symmetric in the dorsal and ventral margins until the furrow is initiated in third instar stage. Loss of *dpp* in the early stage leads to complete failure of eye development (Wharton et al., 1996; Chanut and Heberlein, 1997). Hence, it is possible that the early and late *dpp* expression may be controlled by distinct regulatory elements of the *dpp* gene; and *dpp^{blk}* mutation may affect a specific regulatory element for late ventral expression during furrow initiation. This possibility is consistent with the relatively normal eye disc development until furrow initiation and the sudden loss of *dpp* RNA in the ventral margin in differentiating *dpp^{blk}* mutant disc.

6.3. Function of Lobe and Serrate in ventral growth

Mutations in the *Lobe* (*L*) gene, which was first isolated several decades ago (Morgan et al., 1925), also result in striking ventral-specific growth defects in the eye (Heberlein et al., 1993). In contrast to *dpp^{blk}* mutation, preferential loss of the ventral region is apparent in early second instar disc even in a hypomorphic *L* mutant. The ventral specificity of the *L* function was shown by clonal analysis using a null mutation (Chern and Choi, 2002). *L* mutant clones in the ventral side cause loss of ventral eye tissue. In contrast, the dorsal clones show well-organized photoreceptor clusters, indicating that *L* is required only in the ventral domain of the eye. Genetic epistasis tests suggest that *L* functions are related to N signaling (Chern and Choi, 2002). Reduction of *L* function suppresses the ventral overgrowth by activated N, whereas the dorsal overgrowth is not significantly affected by the reduced *L* function. This further supports that the growth of early eye disc is controlled asymmetrically in the dorsal and ventral domains. Furthermore, small eyes resulted from loss of N function can be partially rescued by overexpression of *L*. These results suggest that *L* functions downstream to N signaling, mediating

N function either in the same or parallel pathway. *L* encodes a novel cytoplasmic protein that is expressed in both dorsal and ventral eye disc throughout development. Expression of *L* in both dorsal and ventral domains is puzzling since its function is only required for growth of the ventral domain. It is possible that *L* may require a ventral-specific gene(s) to carry out its ventral-specific function as the asymmetric function of *tsh* in the dorsal and ventral domains depends on the presence of domain-specific genes (Section 6.1).

N ligand Ser is preferentially expressed in the ventral region of the eye disc during early larval stages (Cho and Choi, 1998; Dominguez and de Celis, 1998). Genetic analysis using loss- and gain-of-function of *L* suggests that *L* is necessary and sufficient to induce Ser expression in early eye disc (Chern and Choi, 2002). Therefore, *L* function in the ventral domain may be mediated by ventral expression of Ser in early eye disc. Interestingly, regulation of Ser expression by *L* depends on the position in the eye disc: *L* is required for Ser expression in most of the ventral region but not in the posterior medial region. This observation suggests that the DV boundary can form even in the absence of *L* since the midline region remains relatively intact to initiate growth of the disc affecting only the ventral region. Therefore, early eye discs are initially patterned into the dorsal and ventral compartments. However, in terms of growth property, the eye disc may be subdivided into three domains: dorsal, *L*-independent medial subdomains, and *L*-dependent ventral (Chern and Choi, 2002).

Hypomorphic *Ser* alleles show reduced eye size indicating that Ser is required for normal growth of the eye (Speicher et al., 1994; Go et al., 1998). In contrast, clones of *Ser* null mutation do not show obvious defects in retinal differentiation, whereas expression of a dominant negative form of Ser causes severe defects in the eye (Hukriede et al., 1997; Sun and Artavanis-Tsakonas, 1997; Kumar et al., 2001b; Chern and Choi, 2002). The lack of phenotype in *Ser* mutant clones suggests that Ser function may be compensated by another factor, or Ser may be secreted or transendocytosed into neighboring cells as shown in the cell culture system (Klueg and Muskavitch 1999). Overexpression of diffusible Ser^{DN} in early eye disc using *ey-GAL4* results in either preferential loss of ventral eye or loss of the entire eye. Overexpression of Ser^{DN} in random clones generated by the *flp*-out method (Pignoni and Zipursky, 1997) mimics the ventral loss caused by *L* null mutant clones. Similar phenotypes of *L* mutants and Ser^{DN} overexpression are consistent with the hypothesis that *L* acts as an upstream regulator of Ser for growth of the ventral domain (Chern and Choi, 2002).

Recent studies have revealed that the extent of ventral-specific eye loss depends on when *L* or *Ser* function is removed. Strikingly, removal of *L* or *Ser* function during early to late first instar results in almost complete failure of eye development rather than the loss of ventral half. Further analysis indicates that the expression of dorsal selector *pnr* emerges in the eye disc during late first instar stage and the eye disc prior to *pnr* expression is equivalent to the ventral state, and therefore becomes sensitive to the loss of *L* or *ser* function. This suggests that the ground state of the larval eye disc is equivalent to the ventral fate, and *L* and *ser* are essential for survival and/or maintenance of this ventral state (Singh and Choi, 2003).

7. Role of DV compartments in organizing planar polarity

We have discussed how DV patterning in early eye disc organizes growth of eye disc. Another important function of DV domain specification is to regulate the signaling processes involved in patterning of DV planar polarity in the eye. It has been proposed that the morphogenetic furrow is important for not only the assembly of ommatidia but also for organizing the DV polarity (Chanut and Heberlein, 1995; Strutt and Mlodzik, 1995). However, it has been shown that generation of ectopic furrows moving backward from anterior to posterior can revert the AP polarity but not the DV polarity of ommatidia (Ma and Moses, 1995; Reifegerste et al., 1997; Wehrli and Tomlinson, 1998). These studies imply independent regulation of these two axes of planar polarity, which is also consistent with early establishment of DV patterning prior to the initiation of furrow. This section will focus on the role of early DV patterning in organizing long-range polarity signals along the DV axis. The processes involved in reading out these signals are discussed in other chapters by M. Mlodzik and M. Simon.

7.1. Signaling in pole-to-equator gradients

Hair patterns on the insect epidermis have been extensively studied to understand how the cells allocated to a body segment or territory is polarized into specific orientations (Gubb and Garcia-Bellido, 1982; Nubler-Jung et al., 1987; Gubb, 1993). Tissue transplantation experiments in other insects such as the plant feeding bug *Oncopeltus* (Lawrence, 1966) and the blood-sucking bug *Rhodnius* (Stumpf, 1966) have suggested that positional information or polarizing signals are provided in concentration-dependent gradients to which pattern elements respond differently depending on how far they are positioned from the signal source (Lawrence, 1992). Genetic analysis of loss-of-function and gain-of-function of polarity genes identified in *Drosophila* also suggests that the uniform planar polarity in the dorsal and ventral eye fields is regulated by gradients of polarizing signals (Wehrli and Tomlinson, 1998, Zeidler et al., 1999a).

What are the signals for DV ommatidial polarity and how are signal gradients established? It has been shown that non-canonical Wg/Wnt signaling pathway is important for determining planar polarity (Boutros et al., 1998; Shulman et al., 1998; Mlodzik, 1999; Reifegerste and Moses, 1999; Boutros et al., 2000), presumably by transducing the polarity signal presented in a gradient form(s). Therefore, an obvious candidate for polarity signal is Wg itself. Interestingly, local overexpression of Wg in clones of cells by *flp*-out methods results in reversal of the polarity toward the polar side of the clone, and such effect is most potent when the clone is located in the equatorial region. Wg expression in the eye disc is likely to generate a pole-to-equator gradient since it is secreted from the dorsal and ventral poles (margins) of an eye disc. Hence, ectopic expression of Wg in the equatorial region will reshape the Wg gradient with an additional Wg peak in the equatorial region (Ma and Moses, 1995; Wehrli and Tomlinson, 1995, 1998), leading to the potent repolarization. This ectopic expression study suggests that Wg can function as a primary polarizing signal, although it remains to be demonstrated whether Wg is strictly required for polarity patterning. Analysis of Wg loss-of-function has been hampered because Wg is involved in other events in eye development and therefore loss of *wg* causes complex defects in the eye (Wehrli and Tomlinson, 1998).

In a simple model, components involved in Wg/Wnt signal transduction are expected to be required for readout of polarity signals. Therefore, abnormal signal transduction in the signal-receiving cells will cause polarity abnormalities restricted within mutant cell clones. Surprisingly, however, loss-of-function clones of the *arrow* (*arr*) gene shows dramatic nonautonomous polarity reversals toward the equator. Arr, a low density lipoprotein (LDL) receptor family protein, is essential for Wg/Wnt signaling by acting as a co-receptor for Wg/Wnt (Tamai et al., 2000; Wehrli et al., 2000; Pandur and Kuhl, 2001). Similar nonautonomous phenotypes, as seen in *arr* mutant, were also found with mutant clones of *dsh* and *arm* that are downstream to *arr*. These nonautonomous polarity changes were more potent as the position of mutant clones was closer to the polar region. This leads to an interesting proposal that Wg/Wnt signal transduction pathway may also be used to generate an unknown polarity signal, called factor X or secondary signal, which forms an equator-to-pole gradient in the opposite direction to the Wg gradient (Wehrli and Tomlinson, 1998; Blair, 1999).

The proposed dual functions of Wg/Wnt signal transduction pathway also raises a question: How can the same pathway carry out both generation and readout of signal? Further analysis of *dsh* mutant clones in developing eye discs suggests that autonomous and nonautonomous functions of *dsh* are required in distinct cell populations. Clones of *dsh* mutant cells within the eye disc proper show polarity reversals within the clone border. In contrast, clones located in the disc margin juxtaposed to the disc proper and peripodial membrane cause nonautonomous long-range polarity reversals. This suggests that *dsh* and *arm* are required in the disc margin for organizing long-range polarity signaling, whereas *dsh* in photoreceptors in the disc proper is necessary only for readout of polarity signal (Lim and Choi, 2004).

7.2. Signaling in equator-to-pole gradients

In addition to a pole-to-equator gradient of Wg (Wehrli and Tomlinson, 1998), another secreted molecule, Unpaired (Upd), has been proposed to be involved in polarity signaling. Upd, the ligand for JAK-STAT signaling pathway (Harrison et al., 1998), is expressed preferentially in the posterior margin of second instar eye disc and persisted at the DV midline adjacent to the optic stalk in third instar disc (Zeidler et al., 1999a). The localization of Upd in the posterior DV border region suggests that Upd forms a equator–polar gradient opposite to the Wg gradient and therefore may be involved in secondary signaling for planar polarity. Loss of function clones of Upd mutation cause position-dependent phenotypes as expected from the localized expression of Upd at the posterior margin. Upd mutant clones located away from the equator show little effect, whereas the clones in the ventral side near the equator cause striking expansion of *mirr* expression, that is, ventral shifting of the equator (Zeidler et al., 1999a). These results indicate that Upd and Wg are important for positioning of the equator by regulating *mirr* expression. In contrast to the equator-to-pole gradient of Upd, *STAT-lacZ* used as a reporter of Upd-JAK activity shows high level expression in the DV polar regions and the lowest level near the equator, indicating a negative regulation of STAT by Upd-JAK (Zeidler et al, 1999a). However, loss of STAT92E does not result in polarity defects as seen in *upd* or *hopscotch (hop, JAK)* (Binari and Perrimon, 1994) mutant clones. This suggests that STAT function may be redundant (Harrison et al., 1998).

Since the gradients formed by secreted Wg and Upd probably have opposite shapes, it is possible that Wg may antagonize Upd expression. However, further tests suggest that Wg and Upd may function independently. First, loss-of-function clones of *hop*, which encodes JAK (Binari and Perrimon, 1994), can induce nonautonomous reversals of planar polarity in the polar side of the clone, but no changes were detected in *mirr-lacZ* expression. Second, overexpression of Wg near the site of Upd expression or Upd overexpression did not affect the expression of the other, respectively (Luo et al., 1999; Zeidler et al., 1999a). Based on this independent regulation of two pathways, it has been proposed that the Upd and Wg may act in parallel to regulate the concentration gradient of a single second signal. Upd may activate this signal at the equatorial region, whereas Wg represses it at the polar region. This will increase the slope of the equator-to-polar gradient of a second signal, thereby allowing more efficient readout of the polarity signal (Zeidler et al., 1999a).

In contrast to the localized expression of *upd* in the equatorial posterior margin, *four-jointed (fj)* is expressed in a broad equatorial domain (Brodsky and Steller, 1996). It is localized in the broad anterior region in contrast to the posterior equatorial expression of *WR122* (Heberlein et al., 1998). Fj is involved in proximo-distal patterning of leg (Villano and Katz, 1995; Zeidler et al., 1999b; Buckles et al., 2001). It encodes a type II-transmembrane protein whose C-terminus is extracellular. Clonal analysis in the leg disc has shown that Fj acts nonautonomously. Biochemical evidence also suggests that the C-terminal extracellular domain is cleaved and secreted to act as a diffusible signaling molecule (Villano and Katz, 1995). Existing Fj alleles including a molecular null mutation are viable, and homozygous mutants show very weak DV reversals. However, patches of mutant clones surrounded by wild-type cells caused strong nonautonomous polarity reversals: Wild-type ommatidia in the polar side of a clone showed reversed polarity and the strength of polarity reversals was enhanced as the clones are located closer to the equator. This suggests that the equator-to-pole gradient of Fj provides an important polarity cue.

While Upd is independent of Wg, ectopic expression of Wg results in downregulation of Fj, suggesting that equatorial expression of Fj is due to negative regulation by Wg at the DV polar regions. Besides these secreted molecules, two cadherin superfamily transmembrane proteins Dachsous (Ds) and Fat (Ft) are important for modulation of Fz activity in R3/4 cells to determine planar polarity (Rawls et al., 2002; Yang et al., 2002; Fanto et al., 2003). Interestingly, Wg not only inhibits Fj expression in the marginal regions (Zeidler et al., 1999a) but also activates the expression of Ds, resulting in a pole-to-equator gradient of Ds (Yang et al., 2002).

Further, Fj enriched in the equatorial region plays a role in inducing asymmetry of Ds and Ft in the R3/4 equivalent precursor pair, resulting in the higher Ds in the R4 polar cells and the higher Ft and Fz function in the R3 equatorial cells (Strutt and Strutt, 2000b; Yang et al., 2002). More detailed analysis of Fj, Ds, and Ft functions in controlling R3/R4 asymmetry is described in the chapter by M. Simon.

In contrast to the Wg-dependent downregulation of Fj, N and Upd act as positive regulators of Fj (Zeidler et al., 1999b; Buckles et al., 2001). Ectopic clones expressing intracellular domain of N (N^{intra}) activates Fj expression in the polar side of the clone (Papayannopoulos et al., 1998). Studies on Wg, Upd, and Fj raise a possibility that long-range polarity signals are generated by a combination of multiple gradients of secreted molecules that may in part be functionally redundant. In this model (Fig. 8), the initial DV patterning is established by compartmentalization of DV domains by domain-specific selector genes such as *pnr* and *Iro-C*. This early DV patterning induces activation of N signaling at the DV border for growth of eye disc and organizing planar polarity signaling in subsequent stages. Polarity signals may consist of multiple gradients of signaling molecules. N signaling from the DV boundary leads to the expression of Upd and Fj in the equator-to-pole gradients. The pole-to-equator gradient of Wg formed by its secretion from the dorsoventral poles is involved in expression of Ds and downregulation of Fj in the polar regions to reinforce the equator-to-pole gradient of Fj. Therefore, Fj and

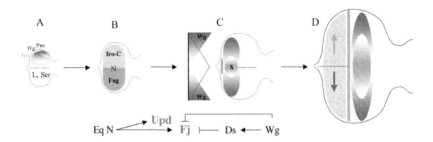

Fig. 8. Schematic presentation of DV patterning events in eye disc. The top panel shows a simplified model of gene functions involved in DV patterning during larval eye development. (A) Mid-late first instar stage. Pnr expressed from the dorsal peripodial cells induces Wg expression. Secreted *wg* allows expression of *Iro-C* genes in the broad dorsal domain. L and Ser are required for growth of the ventral domain. (B) Late first- to mid-second instar stage. Iro-C and Fng domains are established and N is activated at the DV border. (C) Early third instar stage. Wg secreted from the DV polar regions generate pole-to-equator gradient. Opposite equator-to-pole gradients of secreted signaling molecules such as Upd and Fj may be formed from the equatorial region by N activity at the DV border. The equator-to-pole gradient of Fj may be reinforced Ds, a Wg downstream component. Upd and Fj are involved in generation of secondary polarity signal X to which differentiating photoreceptor clusters respond by activating a signal transduction pathway to establish DV planar polarity (step D). (See Color Insert.)

Upd, together or in parallel, are candidates for the secondary signal or are involved in producing a secondary signal.

8. Concluding remarks

DV planar polarity of photoreceptor clusters is established by a number of signaling events throughout eye disc development. We described an overview of key developmental steps involved in early DV patterning and the organization of polarity signal(s). Among a cascade of events, the DV compartmentalization is a key to initiate patterning and growth in early eye disc. Domain-specific genes provide spatial cues for DV poles and midline. These cues are used to induce expression of diffusible molecules that form gradients of polarity signals in the pole-to-equator direction and/or vice versa. Photoreceptor clusters that emerge from the furrow transduce these signals to polarize the cluster pattern into either dorsal or ventral form depending on their position in the eye disc.

Although our knowledge on the DV patterning in the eye has dramatically increased in recent years, there are many unanswered questions in three major areas. The first area is about the regulation of DV domain-specific genes. It is unclear at the molecular level how Pnr expression localized in the dorsal margin of peripodial membrane leads to the expression of *Iro-C* genes in the dorsal half of an eye disc. It is also unknown how *fng* expression is induced and how repression of *fng* by Iro-C in the dorsal domain is derepressed in later stages in the continued presence of *wg* and Iro-C in the dorsal side. The second area is to understand how domain-specific genes and N signaling at the DV boundary contribute to growth of the eye disc. It will be important to understand the molecular basis of asymmetric control of disc growth by a group of genes including *L* and *Ser* as these genes are not only essential for eye morphogenesis but also may be important for organizing the eye pattern. The third area is the relationship between the early DV patterning and the organization of planar polarity. Key issues involve regulatory relationships between N signaling at the DV boundary and the equatorial signals such as Fj and Upd, and the identification of secondary signal(s).

Recent studies in vertebrate visual systems have identified several genes that are expressed in a DV domain-specific pattern in the retina and/or the tectum. BMP-4 and Tbx5 act like dorsal selectors and restrict the expression of Vax2 and Pax2 to the ventral eye (Koshiba-Takeuchi et al., 2000; Mui et al., 2002; Peters, 2002). These DV expression domains comprise developmental compartments (Peters, 2002; Peters and Cepko, 2002) and are important for the regulation of the DV retina-tectal projection pattern (Koshiba-Takeuchi et al., 2000, McLaughlin et al., 2003). Further, Jagged-1, a vertebrate homolog of Ser shows an asymmetric DV expression pattern in the retina and its loss of function results in strong eye reduction (Xue et al., 1999). Therefore, the DV boundary may play conserved roles in organizing growth and pattern of visual system in higher animals, and studies in *Drosophila* may provide valuable insights into the mechanisms of DV patterning genes in vertebrate systems.

Acknowledgments

We wish to thank Helen McNeil, Marek Mlodzik, and members of our laboratory for invaluable comments. We apologize to many investigators whose work might not have been cited due to space constraint or our ignorance. This work was supported by the National Institute of Health.

References

Adler, P.N. 2002. Planar signaling and morphogenesis in *Drosophila*. Dev. Cell 2, 525–535.

Agnes, F., Suzanne, M., Noselli, S. 1999. The *Drosophila* JNK pathway controls the morphogenesis of imaginal discs during metamorphosis. Development 126, 5453–5462.

Alton, A.K., Fechtel, K., Kopczynski, C.C., Shepard, S.B., Kooh, P.J., Muskavitch, M.A. 1989. Molecular genetics of *Delta*, a locus required for ectodermal differentiation in *Drosophila*. Dev. Genet. 10, 261–272.

Bachmann, A., Knust, E. 1998. Positive and negative control of Serrate expression during early development of the *Drosophila* wing. Mech. Dev. 76, 67–78.

Baker, W.K. 1967. A clonal system of differential gene activity in *Drosophila*. Dev. Biol. 16, 1–17.

Baker, W.K. 1978. A clonal analysis reveals early developmental restrictions in the *Drosophila* head. Dev. Biol. 62, 447–463.

Baonza, A., Garcia-Bellido, A. 2000. Notch signaling directly controls cell proliferation in the *Drosophila* wing disc. Proc. Natl. Acad. Sci. USA 97, 2609–2614.

Bessa, J., Gebelein, B., Pichaud, F., Casares, F., Mann, R.S. 2002. Combinatorial control of *Drosophila* eye development by *eyeless*, *homothorax*, and *teashirt*. Genes Dev. 16, 2415–2427.

Bier, E., Vaessin, H., Shepherd, S., Lee, K., McCall, K., Barbel, S., Ackerman, L., Carretto, R., Uemura, T., Grell, E. 1989. Searching for pattern and mutation in the *Drosophila* genome with a P-lacZ vector. Genes Dev. 3, 1273–1287.

Binari, R., Perrimon, N. 1994. Stripe-specific regulation of pair-rule genes by hopscotch, a putative Jak family tyrosine kinase in *Drosophila*. Genes Dev. 8, 300–312.

Blair, S.S. 1995. Compartments and appendage development in *Drosophila*. Bioessays 17, 299–309.

Blair, S.S. 1999. Eye development: Notch lends a handedness. Curr. Biol. 9, R356–R360.

Bosse, A., Zulch, A., Becker, M.B., Torres, M., Gomez-Skarmeta, J.L., Modolell, J., Gruss, P. 1997. Identification of the vertebrate Iroquois homeobox gene family with overlapping expression during early development of the nervous system. Mech. Dev. 69, 169–181.

Boutros, M., Mihaly, J., Bouwmeester, T., Mlodzik, M. 2000. Signaling specificity by Frizzled receptors in *Drosophila*. Science 288, 1825–1828.

Boutros, M., Paricio, N., Strutt, D.I., Mlodzik, M. 1998. Disheveled activates JNK and discriminates between JNK pathways in planar polarity and *wingless* signaling. Cell 94, 109–118.

Brand, A.H., Perrimon, N. 1993. Targeted gene expression as a means of altering cell fates and generating dominant phenotypes. Development 118, 401–415.

Brodsky, M.H., Steller, H. 1996. Positional information along the dorsal-ventral axis of the *Drosophila* eye: Graded expression of the *four-jointed* gene. Dev. Biol. 173, 428–446.

Brower, D.L. 1986. *engrailed* gene expression in *Drosophila* imaginal discs. EMBO J. 5, 2649–2656.

Buckles, G.R., Rauskolb, C., Villano, J.L., Katz, F.N. 2001. *Four-jointed* interacts with *dachs*, *abelson* and *enabled* and feeds back onto the Notch pathway to affect growth and segmentation in the *Drosophila* leg. Development 128, 3533–3542.

Cadigan, K.M., Jou, A.D., Nusse, R. 2002. Wingless blocks bristle formation and morphogenetic furrow progression in the eye through repression of Daughterless. Development 129, 3393–3402.

Calleja, M., Herranz, H., Estella, C., Casal, J., Lawrence, P., Simpson, P., Morata, G. 2000. Generation of medial and lateral dorsal body domains by the *pannier* gene of *Drosophila*. Development 127, 3971–3980.

Calleja, M., Moreno, E., Pelaz, S., Morata, G. 1996. Visualization of gene expression in living adult *Drosophila*. Science 274, 252–255.

Campos-Ortega, J.A., Waitz, M. 1978. Cell clones and pattern formation: Developmental restrictions in the compound eye of *Drosophila*. Dev. Biol. 184, 155–170.

Cavodeassi, F., Diez Del Corral, R., Campuzano, S., Dominguez, M. 1999. Compartments and organising boundaries in the *Drosophila* eye: The role of the homeodomain Iroquois proteins. Development 126, 4933–4942.

Cavodeassi, F., Modolell, J., Campuzano, S. 2000. The *Iroquois* homeobox genes function as dorsal selectors in the *Drosophila* head. Development 127, 1921–1929.

Cavodeassi, F., Modolell, J., Gomez-Skarmeta, J.L. 2001. The *Iroquois* family of genes: From body building to neural patterning. Development 128, 2847–2855.

Chanut, F., Heberlein, U. 1995. Role of the morphogenetic furrow in establishing polarity in the *Drosophila* eye. Development 121, 4085–4094.

Chanut, F., Heberlein, U. 1997. Retinal morphogenesis in *Drosophila*: Hints from an eye-specific *decapentaplegic* allele. Dev. Genet. 20, 197–207.

Chern, J.J., Choi, K.W. 2002. Lobe mediates Notch signaling to control domain-specific growth in the *Drosophila* eye disc. Development 129, 4005–4013.

Cho, K.O., Chern, J., Izaddoost, S., Choi, K.W. 2000. Novel signaling from the peripodial membrane is essential for eye disc patterning in *Drosophila*. Cell 103, 331–342.

Cho, K.O., Choi, K.W. 1998. Fringe is essential for mirror symmetry and morphogenesis in the *Drosophila* eye. Nature 396, 272–276.

Choi, K.W., Benzer, S. 1994. Rotation of photoreceptor clusters in the developing *Drosophila* eye requires the *nemo* gene. Cell 78, 125–136.

Choi, K.W., Mozer, B., Benzer, S. 1996. Independent determination of symmetry and polarity in the *Drosophila* eye. Proc. Natl. Acad. Sci. USA 93, 5737–5741.

Cohen, B., McGuffin, M.E., Pfeifle, C., Segal, D., Cohen, S.M. 1992. *Apterous*, a gene required for imaginal disc development in *Drosophila* encodes a member of the LIM family of developmental regulatory proteins. Genes Dev. 6, 715–729.

Cohen, S.M. 1993. Imaginal disc development. In: *Drosophila Development* (A. Martinez-Arias, M. Bate, Eds.), Cold Spring Harbor: Cold Spring Harbor Press, pp. 747–841.

Crick, F.H.C., Lawrence, P.A. 1975. Compartments and polyclones in insect development. Science 189, 340–347.

Cubadda, Y., Heitzler, P., Ray, R.P., Bourois, M., Ramain, P., Gelbart, W., Simpson, P., Haenlin, M. 1997. *u-shaped* encodes a zinc finger protein that regulates the proneural gene *achaete* and *scute* during the formation of bristles in *Drosophila*. Genes Dev. 11, 3083–3095.

de Celis, J.F., Garcia-Bellido, A., Bray, S.J. 1996. Activation and function of Notch at the dorsal-ventral boundary of the wing imaginal disc. Development 122, 359–369.

Diaz-Benjumea, F.J., Cohen, S.M. 1995. Serrate signals through Notch to establish a Wingless-dependent organizer at the dorsal/ventral compartment boundary of the *Drosophila* wing. Development 121, 4215–4225.

Dietrich, W. 1909. Die Facettenaugen der Dipteren. Z. Wiss. Zool. 92, 465–539.

Diez del Corral, R., Aroca, P., Gomez-Skarmeta, J.L., Cavodeassi, F., Modolell, J. 1999. The Iroquois homeodomain proteins are required to specify body wall identity in *Drosophila*. Genes Dev. 13, 1754–1761.

Doherty, D., Feger, G., Younger-Shepherd, S., Jan, L.Y., Jan, Y.N. 1996. Delta is a ventral to dorsal signal complementary to Serrate, another Notch ligand, in *Drosophila* wing formation. Genes Dev. 10, 421–434.

Dominguez, M., de Celis, J.F. 1998. A dorsal/ventral boundary established by Notch controls growth and polarity in the *Drosophila* eye. Nature 396, 276–278.

Erkner, A., Gallet, A., Angelats, C., Fasano, L., Kerridge, S. 1999. The role of Teashirt in proximal leg development in *Drosophila*: Ectopic Teashirt expression reveals different cell behaviours in ventral and dorsal domains. Dev. Biol. 215, 221–232.

Fanto, M., Clayton, L., Meredith, J., Hardiman, K., Charroux, B., Kerridge, S., McNeill, H. 2003. The tumor-suppressor and cell adhesion molecule Fat controls planar polarity via physical interactions with Atrophin, a transcriptional co-repressor. Development 130, 763–774.

Fanto, M., Mlodzik, M. 1999. Asymmetric Notch activation specifies photoreceptors R3 and R4 and planar polarity in the *Drosophila* eye. Nature 397, 523–526.

Fasano, L., Roder, L., Core, N., Alexandre, E., Vola, C., Jacq, B., Kerridge, S. 1991. The gene *teashirt* is required for the development of *Drosophila* embryonic trunk segments and encodes a protein with widely spaced zinc finger motifs. Cell 64, 63–79.

Fleming, R.J., Gu, Y., Hukriede, N.A. 1997. Serrate-mediated activation of Notch is specifically blocked by the product of the gene *fringe* in the dorsal compartment of the *Drosophila* wing imaginal disc. Development 124, 2973–2981.

Fossett, N., Tevosian, S.G., Gajewski, K., Zhang, Q., Orkin, S.H., Schulz, R.A. 2001. The Friend of GATA proteins U-shaped, FOG-1, and FOG-2 function as negative regulators of blood, heart, and eye development in *Drosophila*. Proc. Natl. Acad. Sci. USA 98, 7342–7347.

Fristrom, D., Fristrom, J. 1993. The metamorphic development of the adult epidermis. In: *The Development of* Drosophila melanogaster (A. Martinez-Arias, M. Bake, Eds.). Cold Spring Harbor: Cold Spring Harbor Laboratory Press, pp. 843–897.

Garcia-Bellido, A., Ripoll, P., Morata, G. 1973. Developmental compartmentalisation of the wing disk of *Drosophila*. Nat. New Biol. 245, 251–253.

Garcia-Bellido, A., Ripoll, P., Morata, G. 1976. Developmental compartmentalization in the dorsal mesothoracic disc of *Drosophila*. Dev. Biol. 48, 132–147.

Gibson, M.C., Lehman, D.A., Schubiger, G. 2002. Lumenal transmission of *decapentaplegic* in *Drosophila* imaginal discs. Dev. Cell 3, 451–460.

Gibson, M.C., Schubiger, G. 2000. Peripodial cells regulate proliferation and patterning of *Drosophila* imaginal discs. Cell 103, 343–350.

Go, M.J., Eastman, D.S., Artavanis-Tsakonas, S. 1998. Cell proliferation control by N signaling in *Drosophila* development. Development 125, 2031–2040.

Gomez-Skarmeta, J.L., Diez del Corral, R., de la Calle-Mustienes, E., Ferre-Marco, D., Modolell, J. 1996. *araucan* and *caupolican*, two members of the novel Iroquois complex, encode homeoproteins that control proneural and vein-forming genes. Cell 85, 95–105.

Gonzalez-Crespo, S., Morata, G. 1995. Control of *Drosophila* adult pattern by *extradenticle*. Development 121, 2117–2125.

Grillenzoni, N., van Helden, J., Dambly-Chaudiere, C., Ghysen, A. 1998. The Iroquois complex controls the somatotopy of *Drosophila* notum mechanosensory projections. Development 125, 3563–3569.

Gubb, D. 1993. Genes controlling cellular polarity in *Drosophila*. Dev. Suppl. 269–277.

Gubb, D., Garcia-Bellido, A. 1982. A genetic analysis of the determination of cuticular polarity during development in *Drosophila melanogaster*. J. Embryol. Exp. Morphol. 68, 37–57.

Haenlin, M., Cubadda, Y., Blondeau, F., Heitzler, P., Lutz, Y., Simpson, P., Ramain, P. 1997. Transcriptional activity of *pannier* is regulated negatively by heterodimerization of the GATA DNA-binding domain with a cofactor encoded by the *u-shaped* gene of *Drosophila*. Genes Dev. 11, 3096–3108.

Haltiwanger, R.S. 2002. Regulation of signal transduction pathways in development by glycosylation. Curr. Opin. Struct. Biol. 12, 593–598.

Harrison, D.A., McCoon, P.E., Binari, R., Gilman, M., Perrimon, N. 1998. *Drosophila unpaired* encodes a secreted protein that activates the JAK signaling pathway. Genes Dev. 12, 3252–3263.

Heberlein, U., Borod, E.R., Chanut, F.A. 1998. Dorsoventral patterning in the *Drosophila* retina by *wingless*. Development 125, 567–577.

Heberlein, U., Wolff, T., Rubin, G.M. 1993. The TGF beta homolog *dpp* and the segment polarity gene *hedgehog* are required for propagation of a morphogenetic wave in the *Drosophila* retina. Cell 75, 913–926.

Heitzler, P., Haenlin, M., Ramain, P., Calleja, M., Simpson, P. 1996. A genetic analysis of *pannier*, a gene necessary for viability of dorsal tissues and bristle positioning in *Drosophila*. Genetics 143, 1271–1286.

Held, L.I., Jr. 2002. The eye disc. In: *Imaginal discs* : Cambridge University Press, pp. 197–236.

Hidalgo, A. 1998. Growth and patterning from the engrailed interface. Int. J. Dev. Biol. 42, 317–324.

Hukriede, N.A., Gu, Y., Fleming, R.J. 1997. A dominant-negative form of Serrate acts as a general antagonist of Notch activation. Development 124, 3427–3437.

Irvine, K.D. 1999. Fringe, Notch, and making developmental boundaries. Curr. Opin. Genet. Dev. 9, 434–441.

Irvine, K.D., Wieschaus, E. 1994. Fringe, a Boundary-specific signaling molecule, mediates interactions between dorsal and ventral cells during *Drosophila* wing development. Cell 79, 595–606.

Ito, K., Awano, W., Suzuki, K., Hiromi, Y., Yamamoto, D. 1997. The *Drosophila* mushroom body is a quadruple structure of clonal units each of which contains a virtually identical set of neurones and glial cells. Development 124, 761–771.

Jaw, T.J., You, L.R., Knoepfler, P.S., Yao, L.C., Pai, C.Y., Tang, C.Y., Chang, L.P., Berthelsen, J., Blasi, F., Kamps, M.P., Sun, Y.H. 2000. Direct interaction of two homeoproteins, homothorax and extra-denticle, is essential for EXD nuclear localization and function. Mech. Dev. 91, 279–291.

Jordan, K.C., Clegg, N.J., Blasi, J.A., Morimoto, A.M., Sen, J., Stein, D., McNeill, H., Deng, W.M., Tworoger, M., Ruohola-Baker, H. 2000. The homeobox gene *mirror* links EGF signaling to embryonic dorso-ventral axis formation through Notch activation. Nat. Genet. 24, 429–433.

Ju, B.G., Jeong, S., Bae, E., Hyun, S., Carroll, S.B., Yim, J., Kim, J. 2000. Fringe forms a complex with Notch. Nature 405, 191–195.

Jurgens, G., Hartenstein, V. 1993. The terminal regions of the body pattern. In: *Drosophila Development* (A. Martinez-Arias, M. Bate, Eds.), Cold Spring Harbor: Cold Spring Harbor Press, pp. 687–746.

Kehl, B.T., Cho, K.-O., Choi, K.-W. 1998. *mirror*, a *Drosophila* homeobox gene in the *Iroquois* complex is required for sensory organ and alular formation. Development 125, 1217–1227.

Kim, J., Irvine, K.D., Carroll, S.B. 1995. Cell recognition, signal induction, and symmetrical gene activation at the dorsal-ventral boundary of the developing *Drosophila* wing. Cell 82, 795–802.

Klein, T., Arias, A.M. 1998. Interactions among Delta, Serrate and Fringe modulate Notch activity during *Drosophila* wing development. Development 125, 2951–2962.

Klueg, K.M., Muskavitch, M.A. 1999. Ligand-receptor interactions and trans-endocytosis of Delta, Serrate and Notch: Members of the Notch signaling pathway in *Drosophila*. J. Cell Sci. 112, 3289–3297.

Koshiba-Takeuchi, K., Takeuchi, J.K., Matsumoto, K., Momose, T., Uno, K., Hoepker, V., Ogura, K., Takahashi, N., Nakamura, H., Yasuda, K., Ogura, T. 2000. Tbx5 and the retinotectum projection. Science 287, 134–137.

Kumar, J.P., Moses, K. 2001a. Eye specification in *Drosophila*: Perspectives and implications. Semin. Cell. Dev. Biol. 12, 469–474.

Kumar, J.P., Moses, K. 2001b. EGF receptor and Notch signaling act upstream of Eyeless/Pax6 to control eye specification. Cell 104, 687–697.

Lawrence, P.A. 1966. Gradients in the insect segment: The orientation of hairs in the milkweed bug *Oncopeltus fasciatus*. J. Exp. Biol. 44, 607–620.

Lawrence, P.A. 1992. The making of a fly. In: *The making of a fly*. Oxford: Blackwell Sci. Pub., pp. 197–236.

Lawrence, N., Klein, T., Brennan, K., Martinez Arias, A. 2000. Structural requirements for Notch signaling with Delta and Serrate during the development and patterning of the wing disc of *Drosophila*. 127, 3185–3195.

Lawrence, P.A., Green, S.M. 1979. Cell lineage in the developing retina of *Drosophila*. Dev. Biol. 71, 142–152.

Lawrence, P.A., Struhl, G., Morata, G. 1979. Bristle patterns and compartment boundaries in the tarsi of *Drosophila*. J. Embryol. Exp. Morphol. 51, 195–208.

Lee, J.D., Treisman, J.E. 2001. The role of Wingless signaling in establishing the anteroposterior and dorsoventral axes of the eye disc. Development 128, 1519–1529.

Lim, J., Choi, K.W. 2004. *Drosophila* eye disc margin is a center for organizing long-range planar polarity. Genesis 39, 26–37.

Luo, H., Asha, H., Kockel, L., Parke, T., Mlodzik, M., Dearolf, C.R. 1999. The *Drosophila* Jak kinase Hopscotch is required for multiple developmental processes in the eye. Dev. Biol. 213, 432–441.

Ma, C., Moses, K. 1995. *wingless* and *patched* are negative regulators of the morphogenetic furrow and can affect tissue polarity in the developing *Drosophila* compound eye. Development 121, 2279–2289.

Masucci, J.D., Miltenberger, R.J., Hoffmann, F.M. 1990. Pattern-specific expression of the *Drosophila decapentaplegic* gene in imaginal disks is regulated by 3' cis-regulatory elements. Genes Dev. 4, 2011–2023.

Maurel-Zaffran, C., Treisman, J.E. 2000. *pannier* acts upstream of *wingless* to direct dorsal eye disc development in *Drosophila*. Development 127, 1007–1016.

McLaughlin, T., Hindges, R., O'Leary, D.D. 2003. Regulation of axial patterning of the retina and its topographic mapping in the brain. Curr. Opin. Neurobiol. 13, 57–69.

McNeill, H., Yang, C.H., Brodsky, M., Ungos, J., Simon, M.A. 1997. *mirror* encodes a novel PBX-class homeoprotein that functions in the definition of the dorsal-ventral border in the *Drosophila* eye. Genes Dev. 11, 1073–1082.

Milner, M.J., Bleasby, A.J., Pyott, A. 1983. The role of the peripodial membrane in the morphogenesis of the eye-antennal disc of *Drosophila melanogaster*. Roux's Arch. Dev. Biol. 192, 164–170.

Mlodzik, M. 1999. Planar polarity in the *Drosophila* eye: A multifaceted view of signaling specificity and cross-talk. EMBO J. 18, 6873–6879.

Moloney, D.J., Panin, V.M., Johnston, S.H., Chen, J., Shao, L., Wilson, R., Wang, Y., Stanley, P., Irvine, K.D., Haltiwanger, R.S., Vogt, T.F. 2000. Fringe is a glycosyltransferase that modifies Notch. Nature 406, 369–375.

Morata, G., Ripoll, P. 1975. Minutes: Mutants of *Drosophila* autonomously affecting cell division rate. Dev. Biol. 42, 211–221.

Morgan, T.H., Bridges, C.B., Sturtevant, A.H. 1925. The genetics of *Drosophila*. Bibliog. Genet. 2, 262.

Mui, S.H., Hindges, R., O'Leary, D.D., Lemke, G., Bertuzzi, S. 2002. The homeodomain protein Vax2 patterns the dorsoventral and nasotemporal axes of the eye. Development 129, 797–804.

Netter, S., Fauvarque, M.O., Diez del Corral, R., Dura, J.M., Coen, D. 1998. *white*[+] transgene insertions presenting a dorsal/ventral pattern define a single cluster of homeobox genes that is silenced by the *Polycomb*-group proteins in *Drosophila melanogaster*. Genetics 149, 257–275.

Neumann, C., Cohen, S. 1997. Morphogens and pattern formation. Bioessays 19, 721–729.

Nubler-Jung, K., Bonitz, R., Sonnenschein, M. 1987. Cell polarity during wound healing in an insect epidermis. Development 100, 163–170.

Okajima, T., Irvine, K.D. 2002. Regulation of Notch signaling by o-linked fucose. Cell 111, 893–904.

Pai, C.Y., Kuo, T.S., Jaw, T.J., Kurant, E., Chen, C.T., Bessarab, D.A., Salzberg, A., Sun, Y.H. 1998. The Homothorax homeoprotein activates the nuclear localization of another homeoprotein, extradenticle, and suppresses eye development in *Drosophila*. Genes Dev. 12, 435–446.

Pan, D., Rubin, G.M. 1998. Targeted expression of *teashirt* induces ectopic eyes in *Drosophila*. Proc. Natl. Acad. Sci. USA 95, 15508–15512.

Pandur, P., Kuhl, M. 2001. An arrow for wingless to take-off. Bioessays 23, 207–210.

Papayannopoulos, V., Tomlinson, A., Panin, V.M., Rauskolb, C., Irvine, K.D. 1998. Dorsal-ventral signaling in the *Drosophila* eye. Science 281, 2031–2034.

Peters, M.A. 2002. Patterning the neural retina. Curr. Opin. Neurobiol. 12, 43–48.

Peters, M.A., Cepko, C.L. 2002. The dorsal-ventral axis of the neural retina is divided into multiple domains of restricted gene expression which exhibit features of lineage compartments. Dev. Biol. 251, 59–73.

Pichaud, F., Casares, F. 2000. *homothorax* and *Iroquois-C* genes are required for the establishment of territories within the developing eye disc. Mech. Dev. 96, 15–25.

Pignoni, F., Zipursky, S.L. 1997. Induction of *Drosophila* eye development by Decapentaplegic. Development 124, 271–278.

Raftery, L.A., Sutherland, D.J. 1999. TGF-beta family signal transduction in *Drosophila* development: From Mad to Smads. Dev. Biol. 210, 251–268.

Ramain, P., Heitzler, P., Haenlin, M., Simpson, P. 1993. *pannier*, a negative regulator of *achaete* and *scute* in *Drosophila*, encodes a zinc finger protein with homology to the vertebrate transcription factor GATA-1. Development 119, 1277–1291.

Rawls, A.S., Guinto, J.B., Wolff, T. 2002. The cadherins Fat and Dachsous regulate dorsal/ventral signaling in the *Drosophila* eye. Curr. Biol. 12, 1021–1026.

Ready, D.F., Hanson, T.E., Benzer, S. 1976. Development of the *Drosophila* retina, a neurocrystalline lattice. Dev. Biol. 53, 217–240.

Reifegerste, R., Ma, C., Moses, K. 1997. A polarity field is established early in the development of the *Drosophila* compound eye. Mech. Dev. 68, 69–79.

Reifegerste, R., Moses, K. 1999. Genetics of epithelial polarity and pattern in the *Drosophila* retina. Bioessays 21, 275–285.

Rieckhof, G.E., Casares, F., Ryoo, H.D., Abu-Shaar, M., Mann, R.S. 1997. Nuclear translocation of Extradenticle requires *homothorax*, which encodes an extradenticle-related homeodomain protein. Cell 91, 171–183.

Shulman, J.M., Perrimon, N., Axelrod, J.D. 1998. Frizzled signaling and the developmental control of cell polarity. Trends Genet. 14, 452–458.

Singh, A., Kango-Singh, M., Choi, K.W., Sun, Y.H. 2004. Dorso-ventral asymmetric functions of *teashirt* in *Drosophila* eye development depend on spatial cues provided by early DV patterning. Mech. Dev. 121, 365–370.

Singh, A., Choi, K.W. 2003. Initial state of the *Drosophila* eye before dorsoventral specification is equivalent to ventral. Development 130, 6351–6360.

Singh, A., Kango-Singh, M., Sun, Y.H. 2002. Eye suppression, a novel function of *teashirt*, requires Wingless signaling. Development 129, 4271–4280.

Speicher, S.A., Thomas, U., Hinz, U., Knust, E. 1994. The *Serrate* locus of *Drosophila* and its role in morphogenesis of the wing imaginal discs: Control of cell proliferation. Development 120, 535–544.

Struhl, G. 1981. A blastoderm fate map of compartments and segments of the *Drosophila* head. Dev. Biol. 84, 386–396.

Strutt, D., Johnson, R., Cooper, K., Bray, S. 2002. Asymmetric localization of Frizzled and the determination of Notch-dependent cell fate in the *Drosophila* eye. Curr. Biol. 12, 813–824.

Strutt, D.I., Mlodzik, M. 1995. Ommatidial polarity in the *Drosophila* eye is determined by the direction of furrow progression and local interactions. Development 121, 4247–4256.

Strutt, H., Strutt, D. 2002a. Nonautonomous planar polarity patterning in *Drosophila*: Disheveled-independent functions of Frizzled. Dev. Cell 6, 851–863.

Strutt, H., Strutt, D. 2002b. Planar polarity: Photoreceptors on a high fat diet. Curr Biol. 12, R384–R385.

Stumpf, H.F. 1966. Mechanism by which cells estimate their location within the body. Nature 212, 430–431.

Sun, X., Artavanis-Tsakonas, S. 1997. Secreted forms of DELTA and SERRATE define antagonists of Notch signaling in *Drosophila*. Development 124, 3439–3448.

Sun, Y.H., Tsai, C.J., Green, M.M., Chao, J.L., Yu, C.T., Jaw, T.J., Yeh, J.Y., Bolshakov, V.N. 1995. *white* as a reporter gene to detect transcriptional silencers specifying position-specific gene expression during *Drosophila melanogaster* eye development. Genetics 141, 1075–1086.

Tamai, K., Semenov, M., Kato, Y., Spokony, R., Liu, C., Katsuyama, Y., Hess, F., Saint-Jeannet, J.P., He, X. 2000. LDL-receptor-related proteins in Wnt signal transduction. Nature 407, 530–535.

Tomlinson, A. 1988. Cellular interactions in the developing *Drosophila* eye. Development 104, 183–193.

Tomlinson, A. 1990. The molecular basis of pattern formation in the developing compound eye of *Drosophila*. Semin. Cell Biol. 1, 229–239.

Tomlinson, A., Strapps, W.R., Heemskerk, J. 1997. Linking Frizzled and Wnt signaling in *Drosophila* development. Development 124, 4515–4521.

Tomlinson, A., Struhl, G. 1999. Decoding vectorial information from a gradient: Sequential roles of the receptors Frizzled and Notch in establishing planar polarity in the *Drosophila* eye. Development 126, 5725–5738.

Tree, D.R., Ma, D., Axelrod, J.D. 2002. A three-tiered mechanism for regulation of planar cell polarity. Semin. Cell Dev. Biol. 13, 217–224.

Treisman, J.E., Rubin, G.M. 1995. *wingless* inhibits morphogenetic furrow movement in the *Drosophila* eye disc. Development 121, 3519–3527.

Tsang, A.P., Visvader, J.E., Turner, C.A., Fujiwara, Y., Yu, C., Weiss, M.J., Crossley, M., Orkin, S.H. 1997. FOG, a multitype zinc finger protein, acts as a cofactor for transcription factor GATA-1 in erythroid and megakaryocytic differentiation. Cell 90, 109–119.

Urbach, R., Technau, G.M. 2003. Segment polarity and DV patterning gene expression reveals segmental organization of the *Drosophila* brain. Development 130, 3607–3620.

Villano, J.L., Katz, F.N. 1995. Four-jointed is required for intermediate growth in the proximal-distal axis in *Drosophila*. Development 121, 2767–2777.

Wehrli, M., Dougan, S.T., Caldwell, K., O'Keefe, L., Schwartz, S., Vaizel-Ohayon, D., Schejter, E., Tomlinson, A., DiNardo, S. 2000. *arrow* encodes an LDL-receptor-related protein essential for Wingless signaling. Nature 407, 527–530.

Wehrli, M., Tomlinson, A. 1995. Epithelial planar polarity in the developing *Drosophila* eye. Development 121, 2451–2459.

Wehrli, M., Tomlinson, A. 1998. Independent regulation of anterior/posterior and equatorial/polar polarity in the *Drosophila* eye; evidence for the involvement of Wnt signaling in the equatorial/polar axis. Development 125, 1421–1432.

Wharton, K., Ray, R.P., Findley, S.D., Duncan, H.E., Gelbart, W.M. 1996. Molecular lesions associated with alleles of *decapentaplegic* identify residues necessary for TGF-beta/BMP cell signaling in *Drosophila melanogaster*. Genetics 142, 493–505.

Wiersdorff, V., Lecuit, T., Cohen, S.M., Mlodzik, M. 1996. Mad acts downstream of Dpp receptors, revealing a differential requirement for *dpp* signaling in initiation and propagation of morphogenesis in the *Drosophila* eye. Development 122, 2153–2162.

Wolff, T., Ready, D.F. 1991. The beginning of pattern formation in the *Drosophila* compound eye: The morphogenetic furrow and the second mitotic wave. Development 113, 841–850.

Wolff, T., Ready, D.F. 1993. Pattern formation in the *Drosophila* retina. In: *Drosophila Development* (A. Martinez-Arias, M. Bate, Eds.), Cold Spring Harbor: Cold Spring Harbor Press, pp. 1277–1325.

Wu, J., Cohen, S.M. 2002. Repression of Teashirt marks the initiation of wing development. Development 129, 2411–2418.

Wu, J.Y., Rao, Y. 1999. Fringe: Defining borders by regulating the Notch pathway. Curr. Opin. Neurobiol. 9, 537–543.

Xue, Y., Gao, X., Lindsell, C.E., Norton, C.R., Chang, B., Hicks, C., Gendron-Maguire, M., Rand, E.B., Weinmaster, G., Gridley, T. 1999. Embryonic lethality and vascular defects in mice lacking the Notch ligand Jagged1. Hum. Mol. Genet. 8, 723–730.

Yang, C.H., Axelrod, J.D., Simon, M.A. 2002. Regulation of Frizzled by Fat-like cadherins during planar polarity signaling in the *Drosophila* compound eye. Cell 108, 675–688.

Yang, C.H., Simon, M.A., McNeill, H. 1999. Mirror controls planar polarity and equator formation through repression of fringe expression and through control of cell affinities. Development 126, 5857–5866.

Zeidler, M.P., Perrimon, N., Strutt, D.I. 1999a. Polarity determination in the *Drosophila* eye: A novel role for Unpaired and JAK/STAT signaling. Genes Dev. 13, 1342–1353.

Zeidler, M.P., Perrimon, N., Strutt, D.I. 1999b. The *four-jointed* gene is required in the *Drosophila* eye for ommatidial polarity specification. Curr. Biol. 9, 1363–1372.

Zhao, D., Woolner, S., Bownes, M. 2000. The Mirror transcription factor links signaling pathways in *Drosophila* oogenesis. Dev. Genes Evol. 210, 449–457.

Recent discoveries in vertebrate non-canonical Wnt signaling: Towards a Wnt signaling network

Petra Pandur

Abt. Biochemie, Universität Ulm, Albert-Einstein-Allee 11, 89081 Ulm, Germany

Contents

Cells that derive from a fertilized egg require a plethora of instructions to coordinately develop a complex organism. The body and organ forming instructions come in the shape of growth factors belonging to a variety of gene families that direct a cell's fate. Fate decisions include processes such as specification, cell polarization, and migration to the appropriate region in the embryo. The Wnt proteins constitute one of the major families of growth factors involved in such decision making in vertebrates and invertebrates. Wnts can signal through the canonical Wnt/β-catenin pathway and through the so-called non-canonical, β-catenin-independent Wnt pathways, the latter challenging us with a whole new set of yet unanswered questions. In

Advances in Developmental Biology
Volume 14 ISSN 1574-3349
DOI: 10.1016/S1574-3349(04)14005-2

this chapter I will summarize the most recent discoveries in the field of vertebrate non-canonical Wnt signaling, which point to the existence of a complex non-canonical Wnt signaling network.

1. Introduction

Wnt proteins constitute a family of secreted glycoproteins that are involved in diverse embryological processes, such as cell differentiation, cell migration, and the establishment of cell polarity (reviewed in Wodarz and Nusse, 1998; Kühl, 2003). These processes are mediated by the activation of different intracellular signaling cascades. The hallmark of the canonical Wnt/β-catenin pathway is the stabilization of cytoplasmic β-catenin due to the Wnt-mediated inhibition of GSK3-β and the subsequent relocalization of β-catenin to the nucleus. There, β-catenin regulates target gene expression by forming a complex with members of the TCF/LEF family of HMG box transcription factors (Fig. 1). Most prominent to this pathway is its extensively studied role in the formation of the dorsal-ventral body axis in vertebrates and its function in carcinogenesis. Members of the Wnt-1/Wg class primarily initiate signaling through β-catenin, which results in axis duplication in *Xenopus* embryos

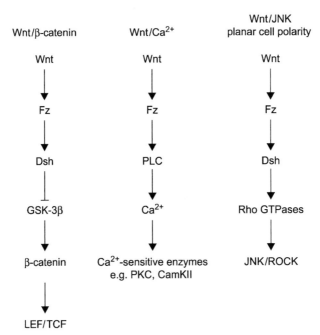

Fig. 1. Simplistic overview of the canonical Wnt/β-catenin pathway and the so far known two non-canonical Wnt pathways that were initially thought to be completely independent of each other. For a detailed description see text and references therein.

and in transformation of C57mg epithelial cells. In contrast, members of the Wnt-5A class activate several β-catenin-independent signaling branches, which have been named the Wnt/Ca^{2+} and the Wnt/JNK signaling pathway, respectively, according to the mediators involved in transducing the incoming Wnt signal. Members of the Wnt-5A class neither induce a secondary body axis, nor transform C57mg cells. Instead they have been implicated in affecting cell movement and antagonizing the canonical Wnt/β-catenin pathway (Du et al., 1995; Torres et al., 1996). Within the past two years detailed studies have now manifested a crucial role for these Wnts in cell migration during gastrulation (Heisenberg et al., 2000; Wallingford et al., 2000; reviewed in Kühl, 2002; Kilian et al., 2003). Just recently, other processes that require cellular polarity, such as neural tube closure and orientation of sensory hair cells in vertebrate ears have also been implicated to be regulated by a vertebrate planar cell polarity-like pathway (Goto and Keller, 2002; Hamblet et al., 2002; Curtin et al., 2003; Dabdoub et al., 2003). Moreover, members of the Wnt-5A class were shown to be involved in very diverse embryological processes such as dorsoventral patterning of the embryo and cardiac induction (Kühl et al., 2000a; Pandur et al., 2002).

To this date we have a very detailed and complex picture of the interactions of the components that make up the Wnt/β-catenin pathway and the reader is referred to according reviews (Wodarz and Nusse, 1998; Huelsken and Behrens, 2002). The functional hierarchy of some of the components of the non-canonical Wnt pathways has been known roughly for a while (Fig. 1). In general it has been accepted that specific Wnt proteins trigger intracellular Ca^{2+} release, which activates the calcium-sensitive protein kinase C (PKC) and Ca^{2+}/calmodulin-dependent kinase II (CamKII), the first two calcium-sensitive proteins identified to be regulated by Wnts (Sheldahl et al., 1999; Kühl et al., 2000a; Kühl, 2002). Downstream targets of this signaling cascade remain largely unknown. The vertebrate Wnt/JNK pathway has been assimilated to represent the planar cell polarity pathway known from *Drosophila*. This pathway has been well established in the fly and the vertebrate homologs of some of its newer components have recently been identified and are under thorough investigation. These components include, for example, the transmembrane protein Strabismus, the LIM-domain containing factor Prickle, and the cadherin-type receptor Flamingo (Darken et al., 2002; Park and Moon, 2002; Morgan et al., 2003; Veeman et al., 2003a). Vertebrate Prickle and Strabismus have been demonstrated to regulate gastrulation movements in zebrafish and *Xenopus*, respectively (Darken et al., 2002; Veeman et al., 2003a). *Flamingo* was shown to play a role in the anteroposterior patterning of the neural plate where it antagonizes the posteriorizing effect of canonical Wnts (Morgan et al., 2003).

Since non-canonical Wnt signaling is a rapidly growing field of research we continually gain a more detailed insight into its growing complexity (reviewed in Strutt, 2003; Veeman et al., 2003b). The focus of this review will be on the most recent findings in the field of vertebrate non-canonical Wnt signaling and I will discuss individual branches that have been investigated in the context of diverse embryological processes emphasizing the biochemical interactions rather than the biological context. Taken together the present data implicates the existence of a complex intertwined non-canonical Wnt signaling nexus.

2. New Wnt receptors in non-canonical Wnt signaling: Ror-2 and Derailed

Still an open question of Wnt signaling is the generation of signaling specificity. Previous data provided the notion that each Wnt ligand and each Frizzled receptor has some preference as to signal through the canonical or a non-canonical Wnt pathway (Du et al., 1995; Kühl et al., 2000b). However, some Frizzleds, for example, X*fz*-7, can function in either pathway (Medina et al., 2000) and specific Wnts can activate different pathways depending on the Frizzled receptor present (He et al., 1997). A key position regarding this question is still held by the intracellular component disheveled (Dsh), which is known to be able to direct an incoming signal to either the β-catenin or the non-canonical pathways through its distinct domains (reviewed in Wharton, 2002). However, it is unclear which input 'instructs' Dsh in how to pass on a particular signal. Recently, the focus on what molecules confer signaling specificity was shifted to the membrane by the finding that signaling through the β-catenin pathway requires a co-receptor of the LDL receptor-like protein (LRP) gene family (Pinson et al., 2000; Tamai et al., 2000; Wehrli et al., 2000). This of course raises the question as to whether Wnts that activate non-canonical signaling pathways also require a co-receptor to do so, a hypothesis that has been implicated by recent publications. Wnt-11 was shown to bind to the atypical tyrosine kinase Ror-2 and the Wnt binding domain in this receptor is required for Ror-2 activity (Hikasa et al., 2002). Dominant-negative Ror-2 mutants block convergent extension movements and this effect can be rescued by constitutively active cdc42 (Hikasa et al., 2002). Interestingly, the tyrosine kinase domain of Ror-2 was not required to elicit this effect. Ror-2 also interacts with Wnt-5A, which leads to JNK activation (Oishi et al., 2003). Additionally, Ror-2-/- and Wnt-5A-/-mice show similar phenotypes (Oishi et al., 2003). Both papers strongly argue for an involvement of Ror-2 in non-canonical Wnt signaling. Supporting the previous finding that Wnt/JNK signaling is involved in heart development (Pandur et al., 2002), mice mutant for Ror-1 and Ror-2 exhibit heart defects (Nomi et al., 2001). Similarly, in *Drosophila,* derailed (drl), an atypical tyrosine kinase of the RYK family binds DWnt-5 and this interaction is required for proper axon guidance in choosing the anterior versus the posterior commissure to cross the midline (Yoshikawa et al., 2003). This effect is specific to the interaction of drl with DWnt-5 since neither Wg nor DWnt-4 could functionally replace DWnt-5. The idea that drl might act as a co-receptor for Frizzled in mediating the DWnt-5 signal could not be substantiated, since neither *fz fz2* double mutants nor mutations in Dsh a downstream *fz* signaling component, elicited the same effect as DWnt-5. This finding opens up the possibility that DWnt-5 may signal through a yet unidentified pathway by binding to drl. Moreover, posterior migration of branchiomotor neurons in zebrafish appears to be mediated through the interaction of Prickle and Strabismus by a novel pathway that may be independent of the non-canonical Wnt/planar cell polarity pathway. This is implied by the findings that neither *slb/wnt-11* or *ppt/wnt-5A* mutations nor the overexpression of a dominant-negative Disheveled that suppresses Wnt-mediated convergent extension movements affect branchiomotor neuron migration (Bingham et al., 2002; Jessen et al., 2002; Carreira-Barbosa et al., 2003). Taken together, these new data indicate that non-canonical Wnt signal reception may be more complex than previously thought and may involve additional Wnt receptor molecules.

3. cGMP: a novel mediator of non-canonical Wnt signaling

An additional signaling branch that seems to be required for Wnt/Ca^{2+} signaling is the formation of cyclic guanosine-3'5'-monophosphate (cGMP), which is a recently described mediator of non-canonical Wnt signaling (Ahumada et al., 2002; reviewed in Wang and Malbon, 2003). Activation of PLC by Wnts involves the β/γ subunits of heterotrimeric G-proteins leaving the Gα subunit behind (Fig. 2). Gα subunits, however, also act as signal transducers suggesting that non-canonical Wnt signaling might involve so far unknown signaling branches. This issue has been investigated recently and has led to the discovery that Wnt-5A/Rfz-2 signaling can modulate the intracellular levels of cGMP (Ahumada et al., 2002). This effect is mediated through regulation of a phosphodiesterase (PDE). In short, Wnt-5A activates PDE through Gαt, which results in a decrease of intracellular cGMP levels. Most interestingly, inhibitors of PDE also interfered with Rfz-2-triggered intracellular calcium release in zebrafish embryos indicating a potential cross talk between Wnt/Ca^{2+} signaling and Wnt/PDE/cGMP signaling. cGMP is known to activate protein kinase G, which phosphorylates and thereby destabilizes β-catenin. Thus, Wnt/PDE/cGMP signaling

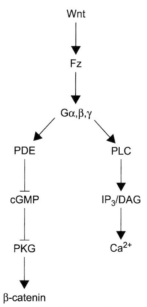

Fig. 2. Wnt-mediated activation of heterotrimeric G-proteins can lead to signaling through cGMP or Ca^{2+} as second messengers. The Gα subunit of G-proteins activates phosphodiesterase (PDE) whereas the Gβ,γ subunits activate phospholipase C (PLC). Low levels of cGMP allow signaling through β-catenin.

appears to interfere with Wnt/β-catenin signaling as well. Initial experiments regarding the biological function of this novel Wnt signaling pathway indicate a requirement for gastrulation movements and primitive endoderm formation (Ahumada et al., 2002). It will be very interesting to further investigate this pathway in a developmental context and integrate it into the network of other signaling pathways.

4. The Wnt/CaCN/NF-AT branch

Central event of the Wnt/Ca^{2+} pathway is the release of intracellular calcium ions, which is initiated by the binding of Wnt-5A or Wnt-11 to specific Frizzled receptors, for example, Rfz-2 (Slusarski et al., 1997b). In addition to PKC and CamKII, a third calcium-sensitive mediator that can be regulated by a Wnt signal has now entered the field, the Ca^{2+}/calmodulin-dependent protein phosphatase calcineurin (CaCN) (Saneyoshi et al., 2002; Fig. 3). A known target of calcineurin is the transcription factor NF-AT (nuclear factor of activated T-cells), which becomes dephosphorylated and subsequently translocates into the nucleus to regulate gene expression. Thus, a well-accepted method to determine calcineurin activation is to monitor the

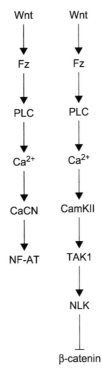

Fig. 3. Functional hierarchy of the Wnt/CaCN/NF-AT branch (left) and the Wnt/CamKII/TAK1/NLK branch (right) as discussed in the text.

translocation of NF-AT into the nucleus. Such translocation of NF-AT takes place after overexpression of XWnt-5A and R*fz*-2 in *Xenopus* animal cap explants (Saneyoshi et al., 2002). Consistent with this observation it was shown that Wnt-5A also promotes nuclear accumulation of NF-AT in T cells in the presence of endothelial cells, a process that has been shown to render T cells resistant to the immunosuppressive drug cyclosporin A (Murphy and Hughes, 2002). Moreover, additional experiments in *Xenopus* link calcineurin/NF-AT signaling in the early embryo to the development of ventral cell fate (Saneyoshi et al., 2002). This is consistent with previous studies implicating IP$_3$ and calcium signaling with ventral development as well (Busa and Gimlich, 1989; Ault et al., 1996; Kume et al., 1997; Kühl et al., 2000a). The ventralizing activity of calcineurin/NF-AT signaling likely results from inhibiting the canonical Wnt/β-catenin pathway downstream of Disheveled and upstream of β-catenin (Saneyoshi et al., 2002). However, the exact molecular mechanism of this interference remains to be determined. It may be that, for example, NF-AT induces the transcription of a negative regulator of β-catenin signaling. Also, there appears to be a cross-talk between canonical and non-canonical Wnt signaling pathways on the level of GSK-3β, since GSK-3β can phosphorylate NF-AT and hence promote its nuclear export (Beals et al., 1997; Ohteki et al., 2000).

Expression of a dominant-negative form of NF-AT in *Xenopus* embryos has a dorsalizing effect leading to the stabilization of β-catenin and consequently to the induction of a secondary axis. Consistent with this is the recent observation that in zebrafish, suppression of an intracellular calcium increase by inhibition of the phosphoinositide (PI) cycle results in expanded nuclear β-catenin domains, and hence in ectopic dorsal gene activation (Westfall et al., 2003).

5. The Wnt/CamKII/TAK/NLK branch

Soon after it had been shown that specific Wnt ligand/Frizzled receptor combinations induce an intracellular increase in calcium concentration, mediators such as CamKII of this new Wnt signaling pathway were characterized. When tested for its embryological activity, CamKII was demonstrated to play a role in determining ventral cell fate (Kühl et al., 2000a). Hence, elevating CamKII activity on the dorsal side of *Xenopus* embryos inhibited dorsal cell fates, whereas reducing CamKII activity on the ventral side promoted dorsal cell fates. Although it is known that Wnt-5A and Wnt-11 can antagonize Wnt/β-catenin signaling, the molecular mechanism of how CamKII could favor ventral development remained obscure. Studies in *Caenorhabditis elegans* and mammalian cells demonstrated that the Wnt/β-catenin pathway can also be regulated by a mitogen-activated protein kinase (MAPK) pathway composed of TAK1 MAPK kinase and nemo-like kinase (NLK) MAPK (Ishitani et al., 1999; Meneghini et al., 1999). TAK1 thereby stimulates NLK activity and subsequently NLK phosphorylates TCF, which prevents the β-catenin/TCF complex from acting as a transcriptional activator (Ishitani et al., 1999). Just recently an upstream signal that regulates the TAK1/NLK cascade has been identified and interestingly this signal is CamKII (Ishitani et al., 2003). Initial assays confirmed that

CamKII interacts with TAK1 and that the CamKII-dependent activation of NLK requires TAK1. Furthermore, CamKII activates NLK in response to an increased intracelluar Ca^{2+} level. This led to the hypothesis that Wnt-5A might antagonize Wnt/β-catenin signaling by activating the TAK1/NLK signaling cascade through CamKII (Fig. 3). Using specific pharmacological inhibitors as well as constitutively active or dominant-negative CamKII mutants this hypothesis was substantiated. Furthermore, this finding confirms previously published data by Kühl et al. (2000a) that CamKII is a component of a signaling pathway that promotes ventral cell fate in *Xenopus* embryos.

Intriguingly, both CamKII and CaCN possess ventralizing activity in the *Xenopus* embryo and in both cases this activity can be regulated by the same ligand, Wnt-5A, however, the molecular mechanisms of their negative interference with the dorsalizing Wnt/β-catenin pathway differ greatly.

There is yet another mechanism that can antagonize β-catenin signaling and this one involves the Gαq pathway of heterotrimeric G proteins (Li and Iyengar, 2002). Using a cell culture system, the authors showed that an intracellular increase in calcium triggered the export of β-catenin from the nucleus to the cytoplasm where β-catenin was degraded by the calcium-sensitive protease calpain. It remains to be determined if this Ca^{2+}-dependent degradation of β-catenin can also be initiated by specific Wnt ligands.

In summary, these findings now demonstrate in a more detailed fashion that the non-canonical Wnt/Ca^{2+} pathway negatively interferes with the canonical Wnt/β-catenin pathway and the mechanisms of interference could not be any more different. Obviously, this opens a new road with a lot of discoveries to come.

6. The Wnt/PKC/Dsh/JNK branch

The existence of a vertebrate planar cell polarity-like pathway, which is also called Wnt/JNK pathway, and its role in gastrulation movements has been fairly well established. Now the focus of research shifts toward detailed analyses of individual components that transduce the incoming Wnt signal in this context. Protein kinase C is probably the most mysterious player in this scenario. PKC comprises a large gene family, whose members differ in the way that they are regulated. Some members are activated by Ca^{2+} and diacylglycerol (DAG), some are only activated by DAG and some are not activated by either Ca^{2+} or DAG. The involvement of PKC in the Wnt/JNK signaling pathway has only recently been implicated in a different study examining the role of Wnt-11/JNK signaling in cardiogenesis (Pandur et al., 2002). With respect to the regulation of convergent extension movements this involvement has now been substantiated and examined in much greater detail. Kinoshita et al. (2003) have identified and characterized a novel PKC isoform in *Xenopus*, PKCδ, which belongs to the group of PKCs that are activated by DAG only. In gain- and loss-of-function studies, PKCδ was shown to be crucial in regulating convergent extension movements in *Xenopus* embryos adding further evidence for the involvement of non-canonical Wnt signaling in gastrulation. More importantly, however, the authors could demonstrate that PKCδ is required for the activation of JNK by Xfz7, a proposed receptor for XWnt-11, in this context. Additionally and not less important, PKCδ interacts with Dsh and is required for recruiting Dsh to the membrane, a prerequisite for Dsh to act in planar polarity signaling. Injections of dominant-negative constructs for PKCα, PKCβ, and PKCδ demonstrated that these PKC isoforms are not interchangeable since only the PKCδ mutant had an effect on gastrulation movements (Kinoshita et al., 2003). Additionally, epistasis experiments in *Xenopus* embryos, place PKCδ upstream of the small Rho GTPase rac and ARP3, a molecule crucial in actin filament assembly and cell motility (Kinoshita et al., 2003). This is consistent with the recent finding that Dsh can activate rac and rho signaling, which is required for *Xenopus* gastrulation (Habas et al., 2003). In detail, it has been demonstrated that the Wnt/fz/Dsh-dependent activation of JNK is mediated by rac but not by rho signaling (Habas et al., 2003). When combined, the so far available data suggests that a major signaling pathway regulating gastrulation movements includes Wnt-11, fz7, PKCδ, Dsh, rac, and JNK.

The Dsh protein comprises three major domains DIX, PDZ, and DEP, the first and the latter of which have been demonstrated to act in Wnt/β-catenin and JNK signaling, respectively. Likewise, activation of rac and rho signaling is also regulated by different domains of Dsh. The central PDZ domain together with the novel Formin homology protein Daam1 (Habas et al., 2001) are required for the activation of rho (Winter et al., 2001), which in turn activates rho kinase. In contrast, JNK activation through rac requires the C-terminal DEP domain of Dsh (Habas et al., 2003) and is independent of Daam-1. Although in this particular study JNK activation was shown to be dependent on rac but not on rho it should be kept in mind that the activation of JNK by Wnts may be regulated in a tissue-specific manner. Indeed, there is evidence that other members of the rho GTPase family, like cdc42 and rhoA, can activate JNK (Tapon and Hall, 1997) and it is known that rho signaling positively regulates JNK activity in *Drosophila* eye development (Mlodzik, 1999). And in yet another study the regulated activation of rho and rac was shown to be required for proper convergent extension movements during *Xenopus* gastrulation (Tahinci and Symes, 2003).

7. Overlapping properties of Wnt/Ca^{2+} and planar cell polarity signaling

Initially, the Wnt/Ca^{2+} pathway and the vertebrate planar cell polarity pathway were both described as pathways regulating cell migration (Slusarski, 1997a; Heisenberg et al., 2000; Tada and Smith, 2000; Wallingford et al., 2000; Kühl et al., 2001). This observation immediately raised the question as to whether these pathways overlap or are even identical or whether both pathways influence cell migration via independent signaling branches. An attempt to answer this question would involve two complementary experimental approaches. It could be tested if components of the Wnt/Ca^{2+} pathway are also involved in a vertebrate planar cell polarity type pathway, and vice versa, it could be tested if components of the planar cell polarity pathway also mediate calcium release. At a glance the findings described above appear to substantiate the first hypothesis, which would be that PKCδ is part of the Wnt/JNK pathway. However, remember that PKCδ is activated by DAG only and does not require elevated concentrations of calcium ions. Thus, PKCδ is not a bona fide component of the Wnt/Ca^{2+} pathway although its activation shares the same upstream component phospholipase C, which subsequently leads to the release of DAG and IP$_3$. The other approach to show an overlap of the Wnt/Ca^{2+} and Wnt/ JNK signaling pathways would be to investigate if components of the Wnt/JNK pathway can regulate components of the Wnt/Ca^{2+} pathway. In fact, this has been reported recently by Sheldahl et al. (2003) who showed that a deletion mutant of Dsh (DshΔDIX) that is known to activate Wnt/JNK signaling (Boutros et al., 1998) also affects the activity of components of the Wnt/Ca^{2+} signaling pathway. In contrast, it does not affect the Wnt/β-catenin pathway. Three independent assays were performed (two of which used *Xenopus* tissues) to verify the function of Dsh in calcium signaling. These included measuring (1) calcium transients in zebrafish embryos, (2) activation of PKCα in enzyme activity assays and by translocation to the membrane, and (3) activation of CamKII in an enzyme activity assay. These data clearly indicate that a molecule that has been proven to be part of the Wnt/JNK signaling pathway is also able to activate Wnt/Ca^{2+} signaling. Furthermore, the authors demonstrated that Dsh-mediated activation of PKC by Wnt-5A and Xfz-7 is pertussis toxin insensitive. Hence, Dsh regulates PKC activity downstream of G-proteins, which are associated with the Frizzled receptors in the membrane. The fact that Dsh can activate PKC might appear confusing at first, since, as discussed above, PKC regulates Dsh activity in gastrulation. However, Dsh was also shown to activate members of the rho GTPase family that can directly activate PLC (Illenberger et al., 1998), which in turn activates PKC. So we end up with a new model for Wnt-mediated PLC activation that is bifurcated downstream of Frizzled with both branches rejoining at PLC (Fig. 4). Although several experimental data argue in favor of such a model it will be necessary to test this in the future. Further evidence for overlapping properties of Wnt/Ca^{2+} and planar cell polarity signaling comes from the identification of the vertebrate homolog of the *Drosophila* gene Prickle that is involved in planar cell polarity signaling in the fly. Veeman et al. (2003a) demonstrated that in addition to its involvement in regulating gastrulation movements and stimulating JNK, Prickle also mediates calcium signaling although it exhibits a different, slower kinetic than do Wnts in this process. How Prickle exerts this effect is unknown and will require further investigation. In summary, the data discussed in this section argue for a partial overlap of Wnt/Ca^{2+} and the planar cell polarity pathway.

8. GSK-3β, mediator of canonical and non-canonical Wnt signaling?

There is emerging evidence for the existence of a non-canonical Wnt signaling pathway that does not signal through β-catenin, however, it does utilize GSK-3β as a mediator. This was shown for example for Wnt-3A and Wnt-7A in neuronal morphogenesis (Hall et al., 2000; Krylova et al., 2002). The molecular nature of this pathway was unclear. However, a recent study implicated the small GTPase cdc42,

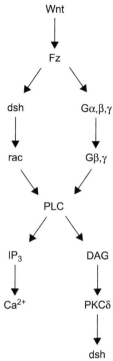

Fig. 4. A proposed mechanism for how a Wnt ligand may activate PLC. Present data suggests that Wnt signal transduction may be diverted at the level of Frizzled and the two different signaling branches rejoin at the level of PLC.

PAR6, and the atypical kinase PKCζ in regulating GSK-3β and APC function independent of β-catenin in establishing cellular polarity (Etienne-Manneville and Hall, 2003). Here, the polarization of the cell depends on the cdc42-mediated phosphorylation of GSK-3β, which leads to an interaction of APC with the plus ends of microtubules to direct cell migration. This pathway is also involved in the establishment of asymmetry during epithelial morphogenesis and in asymmetric cell division, processes all of which involve Wnt signaling (Whangbo et al., 2000; Mizuguchi et al., 2003). As cdc42 activity can be regulated by Wnt signaling it is tempting to speculate that this new signaling branch indeed is activated by Wnt ligands, which surely will be the subject of future investigations.

9. Outlook

Until recently, Wnt proteins were thought to activate individual intracellular signaling pathways (Fig. 1). Recent findings indicate that this is most likely not the case and support the idea of a Wnt signaling network, some aspects of which are summarized in Fig. 5. It remains an exciting challenge to uncover how this network

P. Pandur

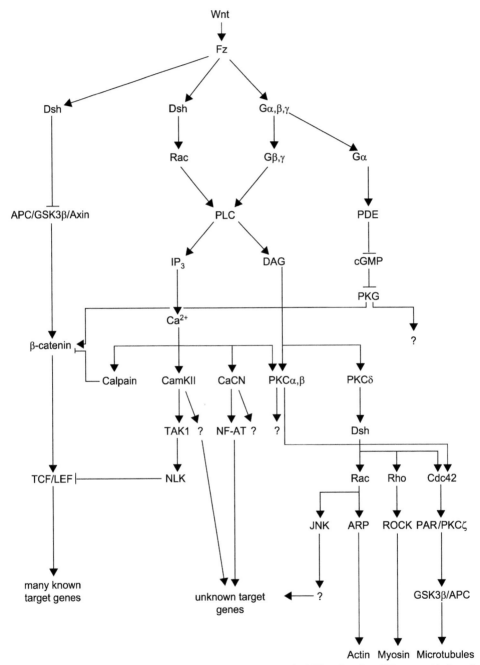

Fig. 5. Summary of the presently available data for non-canonical Wnt signaling. When considering all the interactions of individual components that have been demonstrated to be part in non-canonical Wnt signal transduction cascades a complex network of Wnt signaling emerges that argues against a linear mode of action. Question marks indicate additional so far unknown downstream mediators.

acts *in vivo*, that is, which components are activated in a particular *in vivo* condition and how cell type-specific components of this network determine the outcome of a Wnt signal.

Note Added in Proof

Most recent discoveries add data regarding another complexity of the Wnt signaling network: the signaling specificity (canonical versus non-canonical). Studies of *C. elegans* vulval development suggested that Fz (LIN-17) and Ryk (LIN-18) preferentially bind different Wnt ligands and act in parallel pathways (Inoue et al., 2004). The nature of signal transduction downstream of Ryk (canonical or non-canonical) remains to be determined, however both, LIN-17 and LIN-18, regulate the subcellular localization of TCF (Deshpande et al., 2005). Lu et al. (2004) showed that mammalian Ryk actually functions as a Fz co-receptor and activates the canonical Wnt signaling pathway during the process of neurite outgrowth. In contrast to the study of Ishitani et al. (2003), the Wnt/TAK1 pathway was shown to be activated also by Wnt-1 suggesting that Wnt-1 not only signals through β-catenin but can also inhibit β-catenin-dependent signaling (Smit et al., 2004).

Acknowledgments

I would like to thank Dr. Michael Kühl for helpful discussions and critically reviewing the manuscript.

References

Ahumada, A., Slusarski, D.C., Liu, X., Moon, R.T., Malbon, C.C., Wang, H. 2002. Signaling of rat Frizzled-2 through phosphodiesterase and cyclic GMP. Science 298, 2006–2010.

Ault, K.T., Durmowicz, G., Galione, A., Harger, P.L., Busa, W.B. 1996. Modulation of *Xenopus* embryo mesoderm-specific gene expression and dorsoanterior patterning by receptors that activate the phosphatidylinositol cycle signal transduction pathway. Development 122, 2033–2041.

Beals, C.R., Sheridan, C.M., Turck, C.W., Gardner, P., Crabtree, G.R. 1997. Nuclear export of NF-ATc enhanced by glycogen synthase kinase-3. Science 275, 1930–1934.

Bingham, S., Higashijima, S., Okamoto, H., Chandrasekhar, A. 2002. The zebrafish *trilobite* gene is essential for tangential migration of branchiomotor neurons. Dev. Biol. 242, 149–160.

Boutros, M., Paricio, N., Strutt, D.I., Mlodzik, M. 1998. Disheveled activates JNK and discriminates between JNK pathways in planar polarity and wingless signaling. Cell 94, 109–118.

Busa, W.B., Gimlich, R.L. 1989. Lithium-induced teratogenesis in frog embryos prevented by a polyphosphoinositide cycle intermediate or a diacylglycerol analog. Dev. Biol. 132, 315–324.

Carreira-Barbosa, F., Concha, M.L., Takeuchi, M., Ueno, N., Wilson, S.W., Tada, M. 2003. Prickle 1 regulates cell movements during gastrulation and neuronal migration in zebrafish. Development 130, 4037–4046.

Curtin, J.A., Quint, E., Tsipouri, V., Arkell, R.M., Cattanach, B., Copp, A.J., Henderson, D.J., Spurr, N., Stanier, P., Fisher, E.M., et al. 2003. Mutation of *Celsr1* disrupts planar polarity of inner ear hair cells and causes severe neural tube defects in the mouse. Curr. Biol. 13, 1129–1133.

Dabdoub, A., Donohue, M.J., Brennan, A., Wolf, V., Montcouquiol, M., Sassoon, D.A., Hseih, J.C., Rubin, J.S., Salinas, P.C., Kelley, M.W. 2003. Wnt signaling mediates reorientation of outer hair cell stereociliary bundles in the mammalian cochlea. Development 130, 2375–2384.

Darken, R.S., Scola, A.M., Rakeman, A.S., Das, G., Mlodzik, M., Wilson, P.A. 2002. The planar polarity gene strabismus regulates convergent extension movements in Xenopus. EMBO J. 21, 976–985.

Deshpande, R., Inoue, T., Priess, J.R., Hill, R.J. 2005. lin-17/Frizzled and lin-18 regulate POP-1/TCF-1 localization and cell type specification during C. elegans vulval development. Dev. Biol. 278, 118–129.

Du, S.J., Purcell, S.M., Christian, J.L., McGrew, L.L., Moon, R.T. 1995. Identification of distinct classes and functional domains of Wnts through expression of wild-type and chimeric proteins in Xenopus embryos. Mol. Cell. Biol. 15, 2625–2634.

Etienne-Manneville, S., Hall, A. 2003. Cdc42 regulates GSK-3β and adenomatous polyposis coli to control cell polarity. Nature 421, 753–756.

Goto, T., Keller, R. 2002. The planar cell polarity gene Strabismus regulates convergence and extension and neural fold closure in Xenopus. Dev. Biol. 247, 165–181.

Habas, R., Kato, Y., He, X. 2001. Wnt/Frizzled activation of Rho regulates vertebrate gastrulation and requires a novel Formin homology protein Daam1. Cell 107, 843–854.

Habas, R., Dawid, I.B., He, X. 2003. Coactivation of Rac and Rho by Wnt/Frizzled signaling is required for vertebrate gastrulation. Genes Dev. 17, 295–309.

Hall, A.C., Lucas, F.R., Salinas, P.C. 2000. Axonal remodeling and synaptic differentiation in the cerebellum is regulated by WNT-7a signaling. Cell 100, 525–535.

Hamblet, N.S., Lijam, N., Ruiz-Lozano, P., Wang, J., Yang, Y., Luo, Z., Mei, L., Chien, K.R., Sussman, D.J., Wynshaw-Boris, A. 2002. Disheveled 2 is essential for cardiac outflow tract development, somite segmentation and neural tube closure. Dev. 129, 5827–5838.

He, X., Saint-Jeannet, J.-P., Wang, Y., Nathans, J., Dawid, I., Varmus, H. 1997. A member of the Frizzled protein family mediating axis induction by Wnt-5A. Science 275, 1652–1654.

Heisenberg, C.-P., Tada, M., Rauch, G.-J., Saude, L., Concha, M.L., Geisler, R., Stemple, D.L., Smith, J. C., Wilson, S.W. 2000. Silberblick/Wnt11 mediates convergent extension movements during zebrafish gastrulation. Nature 405, 76–81.

Hikasa, H., Shibata, M., Hiratani, I., Taira, M. 2002. The Xenopus receptor tyrosine kinase Xror2 modulates morphogenetic movements of the axial mesoderm and neuroectoderm via Wnt signaling. Development 129, 5227–5239.

Huelsken, J., Behrens, J. 2002. The Wnt signaling pathway. J. Cell Sci. 115, 3977–3978.

Illenberger, D., Schwald, F., Pimmer, D., Binder, W., Maier, G., Dietrich, A., Gierschik, P. 1998. Stimulation of phospholipase C-beta 2 by the Rho GTPases Cdc42Hs and Rac1. EMBO J. 17, 6241–6249.

Inoue, T., Oz, H.S., Wiland, D., Gharib, S., Deshpande, R., Hill, R.J., Katz, W.S., Sternberg, P.W. 2004. C. elegans LIN-18 is a Ryk ortholog and functions in parallel to LIN-17/Frizzled in Wnt signaling. Cell 118, 795–806.

Ishitani, T., Ninomiya-Tsuji, J., Nagai, S., Nishita, M., Meneghini, M., Barker, N., Waterman, M., Bowerman, B., Clevers, H., Shibuya, H., Matsumoto, K. 1999. The TAK1-NLK-MAPK-related pathway antagonizes signaling between β-catenin and transcription factor TCF. Nature 399, 798–802.

Ishitani, T., Kishida, S., Hyodo, J., Ueno, N., Yasuda, J., Waterman, M., Shibuya, H., Moon, R.T., Ninomiya-Tsuji, J., Matsumoto, K. 2003. The TAK1-NLK Mitogen-Activated Protein Kinase cascade functions in the Wnt-5a/Ca^{2+} pathway to antagonize Wnt/β-catenin signaling. Mol. Cell Biol. 23, 131–139.

Jessen, J.R., Topczewski, J., Bingham, S., Sepich, D.S., Marlow, F., Chandrasekhar, A., Solnica-Krezel, L. 2002. Zebrafish trilobite identifies new roles for Strabismus in gastrulation and neuronal movements. Nat. Cell Biol. 4, 610–615.

Kilian, B., Mansukoski, H., Barbosa, F.C., Ulrich, F., Tada, M., Heisenberg, C.P. 2003. The role of Ppt/Wnt5 in regulating cell shape and movement during zebrafish gastrulation. Mech. Dev. 120, 467–476.

Kinoshita, N., Iioka, H., Miyakoshi, A., Ueno, N. 2003. PKCδ is essential for Disheveled function in a non-canonical Wnt pathway that regulates Xenopus convergent extension movements. Genes Dev. 17, 1663–1676.

Kühl, M., Sheldahl, L.C., Malbon, C.C., Moon, R.T. 2000a. Ca^{2+}/Calmodulin-dependent protein kinase II is stimulated by Wnt and Frizzled homologs and promotes ventral cell fates in *Xenopus*. J. Biol. Chem. 275, 12701–12711.

Kühl, M., Sheldahl, L.C., Park, M., Miller, J.R., Moon, R.T. 2000b. The Wnt/Ca^{2+} pathway a new vertebrate Wnt signaling pathway takes shape. Trends Genet. 16, 279–283.

Kühl, M., Geis, K., Sheldahl, L.C., Pukrop, T., Moon, R.T., Wedlich, D. 2001. Antagonistic regulation of convergent extension movements in *Xenopus* by Wnt/β-catenin and Wnt/Ca^{2+} signaling. Mech. Dev. 106, 61–76.

Kühl, M. 2002. Non-canonical Wnt signaling in *Xenopus*: regulation of axis formation and gastrulation. Sem. Cell Dev. Biol. 13, 243–249.

Kühl, M. 2003. Wnt *signaling in development*, Texas: Landes Biosciences.

Kume, S., Muto, A., Inoue, T., Suga, K., Okano, H., Mikoshiba, K. 1997. Role of Inositol 1,4,5-trisphosphate receptor in ventral signaling in *Xenopus* embryos. Science 278, 1940–1943.

Krylova, O., Herreros, J., Cleverley, K.E., Ehler, E., Henriquez, J.P., Hughes, S.M., Salinas, P.C. 2002. WNT-3, expressed by motoneurons, regulates terminal arborization of neurotrophin-3-responsive spinal sensory neurons. Neuron 35, 1043–1056.

Li, G., Iyengar, R. 2002. Calpain as an effector of the Gq signaling pathway for inhibition of Wnt/β-catenin-regulated cell proliferation. Proc. Natl. Acad. Sci. USA 99, 13254–13259.

Lu, W., Yamamoto, V., Ortega, B., Baltimore, D. 2004. Mammalian Ryk is a Wnt coreceptor required for stimulation of neurite outgrowth. Cell 119, 97–108.

Medina, A., Reintsch, W., Steinbeisser, H. 2000. *Xenopus frizzled 7* can act in canonical and non-canonical Wnt signaling pathways: Implications on early patterning and morphogenesis. Mech. Dev. 92, 227–237.

Meneghini, M.D., Ishitani, T., Carter, J.C., Hisamoto, N., Ninomiya-Tsuji, J., Thorpe, C.J., Hamill, D.R., Matsumoto, K., Bowerman, B. 1999. MAP kinase and Wnt pathways converge to downregulate an HMG-domain repressor in *Caenorhabditis elegans*. Nature 399, 793–797.

Mizuguchi, S., Uyama, T., Kitagawa, H., Nomura, K.H., Dejima, K., Gengyo-Ando, K., Mitani, S., Sugahara, K., Nomura, K. 2003. Chondroitin proteoglycans are involved in cell division of *Caenorhabditis elegans*. Nature 423, 443–448.

Mlodzik, M. 1999. Planar polarity in the *Drosophila* eye: A multifaceted view of signaling specificity and cross-talk. EMBO J. 18, 6873–6879.

Morgan, R., El-Kadi, A.-M., Theokli, C. 2003. Flamingo, a cadherin-type receptor involved in the *Drosophila* planar polarity pathway, can block signaling via the canonical wnt pathway in *Xenopus laevis*. Int. J. Dev. Biol. 47, 245–252.

Murphy, L.L., Hughes, C.C.W. 2002. Endothelial cells stimulate T cell NFAT nuclear translocation in the presence of cyclosporin A: Involvement of the wnt/glycogen synthase kinase-3β pathway. J. Immunol. 169, 3717–3725.

Nomi, M., Oishi, I., Kani, S., Suzuki, H., Matsuda, T., Yoda, A., Kitamura, M., Itoh, K., Takeuchi, S., Takeda, K., Akira, S., Ikeya, M., Takada, S., Minami, Y. 2001. Loss of mRor1 enhances the heart and skeletal abnormalities in mRor2-deficient mice: redundant and pleiotropic functions of mRor1 and mRor2 receptor tyrosine kinases. Mol. Cell. Biol. 21, 8329–8335.

Ohteki, T., Parsons, M., Zakarian, A., Jones, R.G., Nguyen, L.T., Woodgett, J.R., Ohashi, P.S. 2000. Negative regulation of T cell proliferation and interleukin 2 production by the serine threonine kinase GSK-3. J. Exp. Med. 192, 99.

Oishi, I., Suzuki, H., Onishi, N., Takada, R., Kani, S., Ohkawara, B., Koshida, I., Suzuki, K., Yamada, G., Schwabe, G.C., Mundlos, S., Shibuya, H., Takada, S., Minami, Y. 2003. The receptor tyrosine kinase Ror2 is involved in non-canonical Wnt5a/JNK signaling pathway. Genes Cells 8, 645–654.

Pandur, P., Läsche, M., Eisenberg, L.M., Kühl, M. 2002. Wnt-11 activation of a non-canonical Wnt signaling pathway is required for cardiogenesis. Nature 418, 636–641.

Park, M., Moon, R.T. 2002. The planar cell-polarity gene stbm regulates cell behaviour and cell fate in vertebrate embryos. Nat. Cell Biol. 4, 20–25.

Pinson, K.I., Brennan, J., Monkley, S., Avery, B.J., Skarnes, W.C. 2000. An LDL-receptor related protein mediates Wnt signaling in mice. Nature 407, 535–538.

Saneyoshi, T., Kume, S., Amasaki, Y., Mikoshiba, K. 2002. The Wnt/calcium pathway activates NF-AT and promotes ventral cell fate in *Xenopus* embryos. Nature 417, 295–299.

Sheldahl, L.C., Park, M., Malbon, C.C., Moon, R.T. 1999. Protein kinase C is differentially stimulated by Wnt and Frizzled homologs in a G-protein-dependent manner. Curr. Biol. 9, 695–698.

Sheldahl, L.C., Slusarski, D.C., Pandur, P., Miller, J.R., Kühl, M., Moon, R.T. 2003. Disheveled activates Ca^{2+} flux, PKC, and CamKII in vertebrate embryos. J. Cell Biol. 161, 769–777.

Slusarski, D.C., Yang-Snyder, J., Busa, W.B., Moon, R.T. 1997a. Modulation of embryonic intracellular Ca^{2+} signaling by Wnt-5A. Dev. Biol. 182, 114–120.

Slusarski, D.C., Corces, V.G., Moon, R.T. 1997b. Interaction of Wnt and a Frizzled homologue triggers G-protein-linked phosphatidylinositol signaling. Nature 390, 410–413.

Smit, L., Baas, A., Kuipers, J., Korswagen, H., van de Wetering, M., Clevers, H. 2004. Wnt activates the Tak1/Nemo-like kinase pathway. J. Biol. Chem. 279, 17232–17240.

Strutt, D. 2003. Frizzled signaling and cell polarisation in *Drosophila* and vertebrates. Development 130, 4501–4513.

Tada, M., Smith, J.C. 2000. Xwnt-11 is a target of *Xenopus* Brachyury: Regulation of gastrulation movements via Disheveled, but not through the canonical Wnt pathway. Development 127, 2227–2238.

Tahinci, E., Symes, K. 2003. Distinct functions of Rho and Rac are required for convergent extension during *Xenopus* gastrulation. Dev. Biol. 259, 318–335.

Tamai, K., Semenov, M., Kato, Y., Spokony, R., Liu, C., Katsuyama, Y., Hess, F., Saint-Jeannet, J.-P., He, X. 2000. LDL-receptor-related proteins in Wnt signal transduction. Nature 407, 530–535.

Tapon, N., Hall, A. 1997. Rho, Rac and Cdc42 GTPases regulate the organization of the actin cytoskeleton. Curr. Opin. Biol. 9, 86–92.

Torres, M.A., Yang-Snyder, J., Purcell, S.M., Demarais, A.A., McGrew, L.L., Moon, R.T. 1996. Activities of the Wnt-1 class of secreted factors are antagonized by the Wnt-5A class and by a dominant negative cadherin in early *Xenopus* embryo. J. Cell Biol. 133, 1123–1137.

Veeman, M.T., Slusarski, D.C., Kaykas, A., Hallagan Louie, S., Moon, R.T. 2003a. Zebrafish Prickle, a modulator of non-canonical Wnt/*fz* signaling, regulates gastrulation movements. Curr. Biol. 13, 680–685.

Veeman, M.T., Axelrod, J.D., Moon, R.T. 2003b. A second canon: Functions and mechanisms of β-catenin-independent Wnt signaling. Dev. Cell 3, 367–377.

Wallingford, J.B., Rowning, B.A., Vogeli, K.M., Rothbacher, U., Fraser, S.E., Harland, R.M. 2000. Disheveled controls cell polarity during *Xenopus* gastrulation. Nature 405, 81–85.

Wang, H., Malbon, C.C. 2003. Wnt signaling, Ca^{2+}, and cyclic GMP: Visualizing Frizzled functions. Science 300, 1529–1530.

Wehrli, M., Dougan, S.T., Caldwell, K., O'Keefe, L., Schwartz, S., Vaizel-Ohayon, D., Schejter, E., Tomlinson, A., DiNardo, S. 2000. *arrow* encodes an LDL-receptor-related protein essential for Wingless signaling. Nature 407, 527–530.

Westfall, T.A., Hjertos, B., Slusarski, D.C. 2003. Requirement for intracellular calcium modulation in zebrafish dorsal-ventral patterning. Dev. Biol. 259, 380–391.

Whangbo, J., Harris, J., Kenyon, C. 2000. Multiple levels of regulation specify the polarity of an asymmetric cell division in C. elegans. Development 127, 4587–4598.

Wharton, K.A. Jr. 2002. Runnin'with the Dvl: Proteins that associate with Dsh/Dvl and their significance to Wnt signal transduction. Dev. Biol. 253, 1–17.

Winter, C.G., Wang, B., Ballew, A., Royou, A., Karess, R., Axelrod, J.D., Luo, L. 2001. *Drosophila* Rho-associated kinase (Drok) links Frizzled-mediated planar cell polarity signaling to the actin cytoskeleton. Cell 105, 81–91.

Wodarz, A., Nusse, R. 1998. Mechanisms of Wnt signaling in development. Annu. Rev. Cell Dev. Biol. 14, 59–88.

Yoshikawa, S., McKinnon, R.D., Kokel, M., Thomas, J.B. 2003. Wnt-mediated axon guidance via the *Drosophila* Derailed receptor. Nature 422, 583–588.

Planar cell polarity in the vertebrate inner ear

Alain Dabdoub,* Mireille Montcouquiol*
and Matthew W. Kelley

Section on Developmental Neuroscience, National Institute on Deafness and Other Communication Disorders, National Institutes of Health, Bethesda, Maryland 20892

Contents

*Equal contribution, Correspondence to: kelleymt@nidcd.nih.gov.

Advances in Developmental Biology
Volume 14 ISSN 1574-3349
DOI: 10.1016/S1574-3349(04)14006-4

1. General overview

The generation of planar cell polarity (PCP) is a fundamental aspect of embryonic development in all animals. However, until recently, the study of the molecular mechanisms that regulate this process has been restricted to invertebrates, in part because examples of polarized structures are more obvious in these animals. One of the best examples of PCP in vertebrates is the uniform orientation of the stereociliary bundles located on mechanosensory hair cells within the sensory epithelia of the inner ear and lateral line systems. Moreover, hair cell transduction and auditory function are crucially dependent on the development of this planar polarity since stereociliary bundles are only sensitive to vibrations in a single plane. Although the morphological development of uniform bundle orientation was described some time ago, the first manuscripts to identify molecular factors that regulate this process were published in 2003. Not surprisingly, the initial molecules identified as regulators of PCP in vertebrates are closely related to PCP genes in *Drosophila*. However, even from the limited amount of data that is presently available, it appears that there are intriguing differences in the ways that these molecules are used to generate uniform orientation. In this chapter, we will provide an overview of PCP in the auditory system, describe the recent findings regarding molecular control of PCP in vertebrates, and discuss potential similarities and differences between the mechanisms that generate uniform orientation in vertebrates and invertebrates.

2. Vertebrate hair cells

In all vertebrates, sound and motion are perceived through specialized sensory epithelia that are largely restricted to the bilateral inner ears located within the skull adjacent to the hindbrain. Fishes and amphibians have additional similar sensory epithelia located in a lateral line system that runs along the skin of the head and flanks. Each of these sensory epithelia is comprised of specialized mechanosensory cells referred to as sensory hair cells. The number of hair cells per epithelia can vary from as few as ten to twenty in a lateral line neuromast (reviewed in Jones and Corwin, 1993) to over one million in the inner ears of some sharks (Corwin, 1981).

Hair cells, and most other cell types in the inner ear, are derived from ectodermal placodes and are considered to be neuroectodermal in nature (reviewed in Barald and Kelley, 2004). Hair cells are highly polarized with clear differences between their apical (lumenal) and basal-lateral domains. They are always located in the lumenal half of pseudostratified epithelia that also contain an equivalent or greater number of non-sensory cells that are collectively referred to as supporting cells (Fig. 1A). The lumenal surface of each hair cell contains a group of specialized projections, referred to as stereocilia, that protrude into a fluid-filled lumenal space (Fig. 1A). Each individual stereocilium is a modified microvillus that is comprised of a dense core of actin filaments covered by a continuous layer of cellular membrane (Fig. 1B). The number of individual stereocilia per hair cell varies from as few as 30 to 40 to as many as several hundred (reviewed in Hackney and Furness, 1995). However, regardless of the number,

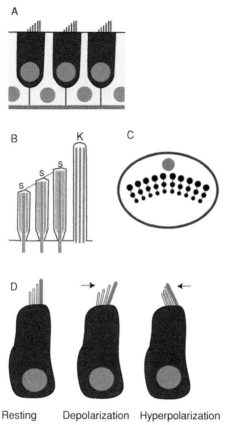

Fig. 1. Anatomy of mechanosensory hair cell epithelia. (A) Cross-section of a typical hair cell sensory epithelium. The epithelium is pseudostratified, with hair cells (blue) located near the lumenal surface and supporting cells (yellow) positioned on the basement membrane (gray). Supporting cells extend lumenal projections that interdigitate between each hair cell. A stereociliary bundle (black and red) extends from the lumenal surface of each hair cell. (B) Cross-section through the center of a stereociliary bundle. The bundle comprises multiple rows of stereocilia (S) and a single kinocilium (K). The core of each stereocilium comprises multiple actin filaments (gray) arranged in parallel. Note that stereocilia narrow at the base to create a flex point. Stereocilia are arranged in a staircase pattern such that the talles stereocilia are located closest to the kinocilium. The top of each stereocilium is connected to the side of the adjacent taller stereocilium by a filamentous strand referred to as a tip link (green). There are other links located on the sides of each stereocilium but these have not been illustrated for clarity. The core of the single kinocilium is comprised of a $9 + 2$ arrangement of microtubules (red) and there are no tip links between the stereocilia and kinocilium. (C) View of a stereociliary bundle at the lumenal surface of a hair cell. Stereocilia are arranged in rows to create a "V" or rounded shape (not shown) with the kinocilium located at the vertex of the bundle. Tip links (green) extend between stereocilia in each row. (D) Illustration of hair cell transduction. With the stereociliary bundle at rest, hair cells maintain a resting negative potential. Deflection of the bundle in the direction of the kinocilium (red) leads to a rapid depolarization of the cell. In contrast, deflection in the opposite direction results in cellular hyperpolarization. Deflections perpendicular to the plane described above have no effect on cellular polarization. (See Color Insert.)

stereocilia are always arranged in a cohesive group that is referred to as a stereociliary bundle. In addition, the lengths of the individual stereocilia within a bundle vary systematically to create a staircase-like shape in which the stereocilia on one side of the bundle are noticeably longer than the stereocilia on the opposite edge (Fig. 1B,C). As a result, all stereociliary bundles have an inherent polarity. In addition, all stereociliary bundles contain a single true cilium that comprises a $9 + 2$ arrangement of microtubules. This cilium, referred to as a kinocilium, is derived from the single true cilium located on all epithelial cells. The kinocilium is always asymmetrically located directly adjacent to the tallest row of stereocilia in a position that correlates with the equator of the bundle (Fig. 1B,C). Kinocilia persist throughout the life of all stereociliary bundles except those on hair cells in the mammalian cochlea, where they are present during initial development but then degenerate with maturity.

The stereociliary bundle acts as the primary transduction apparatus for the mechanosensory signals that form the basis for our perception of sound and movement. Typically, the stereociliary bundle projects into a fluid that is high in potassium and calcium while the hair cell itself maintains comparatively lower levels of both ions. In addition, the lumenal surfaces of hair cells and supporting cells are tightly linked to form an impermeable reticular lamina that prevents ion flow between the lumenal and basolateral surfaces of hair cells. Mechanosensory stimuli generate deflections in the stereociliary bundle that result in the opening of mechanically gated channels located on the tips and edges of each stereocilia. Although these channels have not been characterized at a molecular level, they have been shown to allow passage of calcium and potassium into the cell. The inward flow of ions results in a depolarization of the cell that leads to changes in the rate of neurotransmitter release on the hair cell's basolateral surface (Fig. 1D). Surprisingly, the kinocilium plays no role in transduction, and its degeneration in maturing mammalian cochlear hair cells suggests that its role may be developmental in nature.

The overall sensitivity of the stereociliary bundle is quite high, with deflections of as little as 10 nm resulting in an influx of positive ions. However the bundle is also directionally sensitive. Deflection of the bundle towards the tallest stereocilia results in depolarization of the cell while deflection towards the shortest stereocilia results in a hyperpolarization (Fig. 1D). Interestingly, deflections of the bundle perpendicular to the axis of the staircase result in no change in membrane potential. The basis for this directional sensitivity is the presence of small filamentous structures, referred to as tip links, that span the distance between the tip of one stereocilia and side of an adjacent taller one (Fig. 1B,C). These tip links are believed to physically interact with the mechanically gated ion channels located on each stereocilia such that deflection of the bundle increases the tension on the link, resulting in the generation of a mechanical force that is used to physically open the transduction channel. Since all tip links are oriented in the same plane as the staircase (Fig. 1C), only deflections along this axis will result in opening of the transduction channels.

The precise sensitivity of the stereociliary bundle is reflected in the orientations of hair cells within most sensory epithelia. Many of the sensory epithelia that are associated with perception of motion (vestibular) comprise two populations of hair cells with stereociliary bundles that are oriented along the same plane of polarization but with opposite orientations (Fig. 2). In contrast, in the primary auditory structures in both birds and mammals, all hair cells stereociliary bundles are oriented in the same direction (Figs. 2 and 3). In each case, the overall biophysical structure of the inner ear has evolved such that sound waves generate deflections that are restricted to the plane of stereociliary bundle orientation. As a result, defects in the generation of uniform stereociliary bundle orientation results in auditory (Yoshida and Liberman, 1999) and most probably, vestibular deficits.

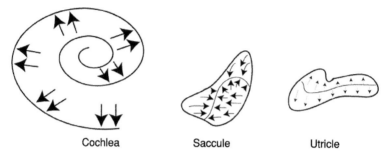

| Cochlea | Saccule | Utricle |

Fig. 2. Planar cell polarity within different hair cell sensory epithelia. The sensory epithelium in the mammalian cochlea extends as a narrow strip of cells along the spiral of the cochlear duct. At all locations hair cells are oriented such that the outward deflections (arrows) of the stereociliary bundle results in depolarization of the cell. In the mammalian saccule, a vestibular sensory structure, hair cells in each half are polarized towards a reversal zone located in the center of the epithelium. In contrast, in a second vestibular epithelium, the utricle, hair cells in each half are polarized towards the outer edge of the structure.

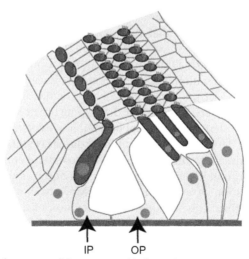

IP   OP

Fig. 3. Three dimensional structure of the sensory epithelium of the mammalian cochlea (also referred to as the organ of Corti). Hair cells are arranged into a single row of inner hair cells (green) and three rows of outer hair cells (blue). Supporting cells (yellow) are specialized as well, with the most striking change being the formation of the tunnel of Corti (TC) a fluid-filled structure that forms between the inner and outer pillar cells (IP and OP). (See Color Insert.)

3. Development of stereociliary bundle orientation

In order to understand how different molecular factors influence the development of uniform stereociliary bundle orientation, it is first necessary to understand how uniform orientation is achieved. Although there have been a limited number of descriptive studies of the developmental process that leads to uniform bundle orientation, a fairly clear picture of the events can be generated. To date, the majority of studies on development of stereociliary bundle orientation have been restricted to the avian cochlea (basilar papilla) and the mammalian cochlea and vestibular sensory patches. The first studies in each of these systems examined the initial orientation of developing stereociliary bundles at the time point at which those bundles became distinct from surrounding microvilli. In all three systems, the initial orientations of most developing stereociliary bundles are non-uniform and show marked deviation from their final orientation. However, as development proceeds, maturing stereociliary bundles gradually obtain uniform orientation. For instance, in the developing mouse cochlea, the orientations of developing stereocilia at the earliest identifiable time points (embryonic day 17 (E17)) deviate from their mature orientation by an average of approximately 70°, while by postnatal day 10 (P10) the average deviation has decreased to approximately 3° (Fig. 4 and Dabdoub et al., 2003). This process has been termed "reorientation" and is presumed to occur as a result of a circumferential movement of the bundle along the outer edge of each hair cell (Fig. 4). However, it is important to consider that another possibility would be the rotation of the entire cell. At present it has not been determined which of these processes actually occurs.

Although the initial orientations of developing stereociliary bundles are non-uniform, results from both auditory and vestibular epithelia indicate that initial orientations are not completely random either. In fact, the initial orientations of most stereociliary bundles at the time that they can first be identified as distinct bundles appear to fall within an arc of 90° that is centered around the final orientation (Cotanche and Corwin, 1991; Denman-Johnson and Forge, 1999; Dabdoub et al., 2003). Therefore the polarization of the developing stereociliary bundle begins prior to the overt differentiation of the bundle. The basis for this initial polarization is still poorly understood; however, studies of the developing organ of Corti indicate that one of the first identifiable steps in hair cell polarization is a centrifugal movement of the single cilium that will develop as the kinocilium from the center of the lumenal surface of the developing hair cell to the periphery (Montcouquiol et al., 2003). The overall direction of this migration is in the general direction of the final orientation; however, there can be a deviation of as much as 180° (McKenzie and Kelley, unpublished observations). Therefore, it is mistakes in the direction of the centrifugal movement that are then corrected during the second, reorientation phase, of bundle polarization.

The mammalian organ of Corti provides a unique opportunity to study the development of uniform bundle orientation over time. Hair cells within the organ of Corti are arrayed in a unique three dimensional pattern that extends along both short (neural-to-abneural) and long (base-to-apex) axes (Figs. 2 and 3). In addition, overall development of all cells within the organ occurs in a gradient that initiates at one edge of the structure and then extends along both the long and short axes (Sher, 1971).

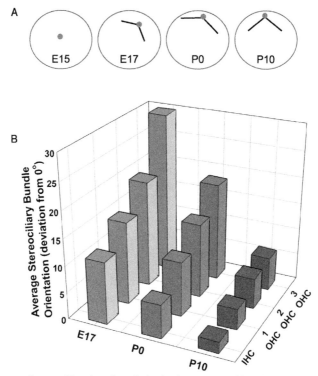

Fig. 4. Development of stereociliary bundle polarity in the mouse cochlae. (A) Prior to bundle formation (E15) each developing hair cell has a single cilium located in the center of the lumenal surface of the cell. As this cilium begins to develop as the kinocilium, it also undergoes a centrifugal migration towards the outer edge of the lumenal surface (E17). The direction of centrifugal migration is non-random but can deviate from the final position by as much as 90°. During this same time period, individual microvilli located on the lumenal surface begin to develop as the stereociliary bundle (black). Following the centrifugal migration, the bundle gradually orients towards its final position through a circumferential movement referred to as reorientation (P0). In its mature form, all bundles are uniformly oriented (P10). (B) Graphic illustration of the change in average deviations in bundle orientations. This graph utilizes the gradient in bundle maturity to illustrate the progressive improvement in bundle orientation that occurs during the reorientation period. Average deviation from 0° on the Z-axis assumes that an orientation of 0° represents the orientation for mature hair cells. The Y-axis illustrates the three time points examined while the X-axis illustrates the hair cell location (see Figure 3 for additional details). Because hair cells develop in a gradient that progresses from inner hair cells (IHC) to first, second, and third row outers. The grid for 3 OHC at E17 represents the most immature bundle while the grid for IHC at P10 represents the most mature bundle.

It is interesting to note that there are marked gradients in the development of stereociliary bundle polarization; however, these gradients do not completely corre-late with the overall developmental gradient. For instance, an analysis of the overall development of bundle orientation indicated a gradient that correlates with the gradient that progresses along both the long and short axes (Dabdoub et al., 2003). However, examination of the initial centrifugal migration indicated that a gradient in the accuracy of this event also exists along the short axis of the organ of Corti. In other words, the deviation in the direction of the centrifugal migration of the developing kinocilium was lowest in inner hair cells, and increased progressively in first, second, and third row outer hair cells (McKenzie and Kelley, unpublished). These results are consistent with the hypothesis that the factors that regulate bundle orientation are located closer to the inner hair cells and polarizing signals are passed sequentially along the short axis from one row of hair cells to the next.

In addition to embryogenesis, the avian auditory system is also capable of generating new hair cells following different kinds of trauma that result in hair cell loss. This regenerative process leads to the formation of new hair cells that recapitulate the embryologic development of these cells, including the generation of uniform stereociliary bundle orientation. As was described for embryologic development, regenerating hair cells arise with a non-random orientation and as the cells differentiate the stereociliary bundles gradually become polarized (Cotanche and Corwin, 1991; Woolley et al., 2001). These new bundles are aligned with respect to the existing bundles in the regions surrounding the site of regeneration. The observation that bundle orientation improves over time during regeneration suggests that polarizing signals persist in the epithelium and that individual hair cells are able to orient their stereociliary bundles based on those signals.

While the preceding sections described the development of bundle polarization as occurring in two unique steps, it is important to consider that these steps may also represent a continuum of progressive polarization, that has been arbitrarily divided based on the ability to identify a developing stereociliary bundle using morphological criteria. At present it is not possible to distinguish between these possibilities. However, there is data (described below) indicating that *wnt-frizzled* signaling plays a role only in the reorientation phase of bundle polarization, suggesting that the separation of polarization into two phases may reflect distinct biological events.

4. Genes that regulate PCP in the mammalian inner ear

Numerous studies have identified a group of genes in *Drosophila*, referred to as the "core PCP genes" which when mutated lead to severe planar polarity defects that affect ommatidial rotation in eyes and bristle orientation in both the wing and thorax (Adler, 2002; Eaton, 2003; Strutt, 2003; Veeman et al., 2003). These genes include the *frizzled* transmembrane receptors (*fz, fz2*), the 7-transmembrane atypical cadherin, *flamingo* (*fmi*), the novel transmembrane protein *van gogh/strabismus* (*vang/stbm*) and the intracellular proteins *disheveled* (*dsh*), *prickle* (*pk*), and *diego*. In mammals, orthologs for most of these genes exist and, in addition, have been duplicated to the point that many exist as gene families. To date, in mammals, there are up to 10 *frizzled* othologs (*fz1* through 10), three *disheveled* (Dvl1-3) othologs, three *flamingo* othologs (*Celsr1-3*), two *vang/stbm* othologs (*vangl1, 2*), and two *prickle* othologs (*pk1, 2*). This multiplication of genes, and the existence of compensation phenomena, has made it difficult to identify genes involved in planar polarity through the generation of single gene deletions. As will be described below, only two orthologs of the core PCP genes in *Drosophila* have been shown to also regulate PCP in mammals. Moreover, our recent results have also identified a role for Scribble 1 (*scrb1*), a mammalian ortholog of the *Drosophila* apical-basal polarity gene Scribble (*scrb*), in PCP (Montcouquiol et al., 2003). *Drosophila scrb* is involved in apical-basal polarity, but does not play an obvious role in PCP (Bilder and Perrimon, 2000).

5. *Vangl2*

In *Drosophila*, *vang/stbm* was independently identified by two separate groups as a gene that causes PCP defects in the wing (*vang*) and the eye (*stbm*) (Taylor et al. 1998; Wolff and Rubin, 1998). *vang/stbm* (hereafter referred to as *vang*) encodes a protein with four putative transmembrane domains and a PDZ domain-binding motif at its carboxy terminus (Wolff and Rubin, 1998). Orthologs for *vang* have been identified in both vertebrates and invertebrates and sequence analysis indicates that the gene has been highly conserved throughout evolution (Wolff and Rubin, 1998; Kibar et al., 2001; Darken et al., 2002; Goto and Keller, 2002; Park and Moon, 2002).

More recently, one of the mammalian orthologs for *vang, vangl2*, has been demonstrated to regulate PCP in the ear (Montcouquiol et al., 2003). Analysis of the cochleae from *Looptail* (*lp*) mice, a spontaneous mouse mutant line, with a point mutation in the *vangl2* gene (Kibar et al., 2001; Murdoch et al., 2001), indicated two major defects in the development of the inner ear: a reduction of the length of the cochlear duct, probably due to a defect in the convergent extension of the duct during development, and a strong disruption of the uniform orientation of stereociliary bundles (Montcouquiol et al., 2003; Fig. 5). Importantly, the overall development of the bundles appeared normal, suggesting that the role of *vangl2* is restricted to determination of orientation. In addition, defects in PCP were not restricted to regions of the cochlea that were noticeably shortened, suggesting that the defect in planar polarity occurs independently of the defect in cochlear extension. Finally, recent results indicate that the defects in uniform bundle polarization are not restricted to the cochlea, and are present in the vestibular system as well (M. Warchol, D. Dickman, D. Huss, M. Montcouquiol, and M.W. Kelley, unpublished data), indicating that *vangl2* is involved in PCP throughout the ear, and probably in many other mammalian systems.

A particularly intriguing aspect of the effects of the *vangl2* mutation in the ears of *lp* mice, was the observation that stereociliary bundle orientation was not equally effected in all hair cells. Differences in the severity and pattern of defect were observed between inner hair cells and outer hair cells, as well as at different positions along the basal-to-apical axis of the cochlea. In particular, inner hair cells and third row outer hair cells were affected more severely than either first row or second row outer hair cells. In addition, while the severity of bundle deviation for outer hair cells was largely invariant along the basal-to-apical axis of the cochlea, inner hair cell deviations varied along the length of the axis with bundles closest to the apex having the worst overall orientation. Moreover, as will be discussed below, a mutation in *scrb1* affected only outer hair cells. Based on these observations, we suggested that different mechanisms might regulate PCP in inner versus outer hair cells, although a molecular basis for these differences has not been determined.

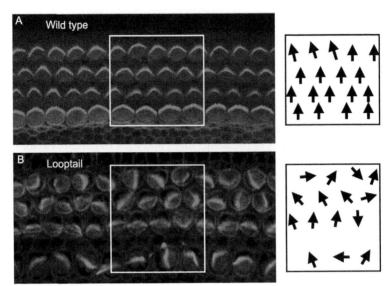

Fig. 5. PCP is disrupted in the cochleae of *lp* mutants. (A) Phalloidin-labeling of hair cell boundaries and stereociliary bundles in a wild-type cochlea at E18. Note that all of the bundles are oriented towards the top edge of the epithelium. Orientations for the hair cells located in the white boxed region are illustrated in the line diagram to the right. With a few exceptions in the third row of outer hair cells, all bundles are uniformly oriented. (B) Stereociliary bundles from the inner ear of a *lp* homozygote. There is a significant disruption in uniform bundle orientation that is clearly illustrated in the line drawing on the right.

The specific functions of the *vang/vangl2* proteins are not clear. Localization studies using both a *vang*-yellow fluorescent protein (*vang*-YFP), fusion protein, and specific antibodies raised against *vang*, indicate that it is expressed in a dynamic pattern in young *Drosophila* wings (Strutt, 2001; Bastock, 2003). In early stages of development, the protein localizes patchily around apicolateral cell boundaries and it co-localizes with armadillo (a β-catenin homologue) at the adherens junction (Fig. 6). With time, an asymmetrical distribution develops, and there is a preferential distribution of *vang* to the proximal boundary of each cell (Strutt, 2001; Bastock, 2003; Fig. 7). Mutation in *vang* results in reduction of apicolateral recruitment of both *fz* and, to a lesser extent, *fmi*. *vang* is also required for *dsh* apicolateral recruitment and for efficient localization of *pk* to membranes. It is believed that, as a first step towards the establishment of PCP, the atypical cadherin *fmi* mediates apicolateral recruitment of planar polarity proteins including *fz* and *vang* to the adherens junction. These three proteins are then required for the recruitment of *dsh, pk,* and *diego*. Following this apical recruitment, an asymmetric localization occurs, and all proteins become asymmetrically distributed along the proximodistal axis of the wing (Bastock, 2003; Fig. 7). This asymmetrical distribution is now considered to be a very characteristic pattern of PCP in *Drosophila* (Usui et al., 1999; Axelrod, 2001; Strutt, 2001; Tree et al., 2002; Bastock et al, 2003). The exact molecular mechanisms leading to this asymmetry are largely unknown, but clearly require the cooperation of all the PCP molecules.

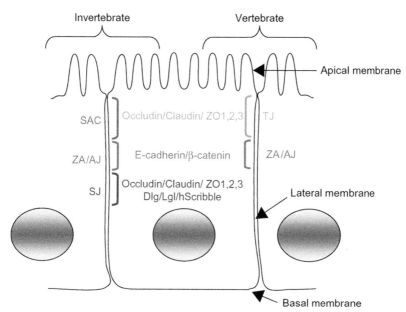

Fig. 6. Comparison of cell junctions in vertebrates and invertebrates. Apical-basal polarity depends on the establishment of cell–cell junctions between the basolateral and apical compartments. In invertebrates, the zonula adherens (also called adherens junction (ZA/AJ)) is located above the septate junction (SJ) while in vertebrates the tight junction (TJ) which is analogous with the SJ is located above the ZA/AJ. Scribble, *dlg*, and *lgl* play key roles in targeting of junctional proteins such as Claudins and ZO proteins to the septate junction in invertebrates. A similar role for *scrb*1 has not been demonstrated yet.

Asymmetric subcellular localization of *vang* protein also occurs in developing photoreceptors during the establishment of ommatidial polarity (Das et al., 2002; Strutt et al., 2002; Yang et al., 2002; Rawls and Wolff, 2003). Overall polarity of the eye is dependent on the specification of individual cells as the R3 and R4 photoreceptors through a mechanism in which *fz* establishes an initial small bias that is subsequently amplified through the *Delta*-Notch pathway (Fig. 7). *fz* signaling specifies R3 and induces *Delta* (a transmembrane ligand) to activate notch receptors in R4. *vang* is required for the R4 fate. *vang* exhibits a dynamic pattern of expression in the R3/R4 pair. In the R3/R4 photoreceptor pair, *vang* preferentially localizes on the R4 side of the common cell–cell boundary while *fz* preferentially localizes on the R3 side of the same boundary (Strutt et al., 2002). Therefore, *vang* and *fz* become localized on opposing cell membranes in R3 and R4, respectively. Based on these observations, the R3/R4 boundary appears to be functionally equivalent to the distal/ proximal cell boundary between cells in the wing.

The specific intracellular role of *vangl2* protein in the generation of uniform bundle orientation is completely unknown. Based on existing data from *Drosophila*, a crucial first step will be to determine the specific subcellular localization of *vangl2* during the development of stereociliary bundles in hair cells, even though, as suggested previously by Strutt (2003), asymmetrical localization of *vangl2* in mammalian cells,

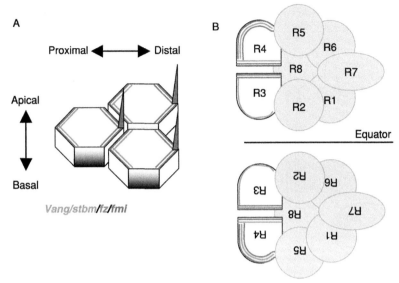

Fig. 7. Localization of core PCP genes during generation of polarity. (A) In the fly wing a homogenous population of cells each generates a single localized hair through differential localization of *vang, stbm, fz,* and *fmi*. While the existence of an actin-based structure of these cells is very similar to hair cells, the systems are dissimilar in that the cells are homogenous. (B) Cellular distribution of the same core PCP genes during development of polarity in the fly eye. The population of varied cell types may be more analogous with the mammalian cochlea and inner ear. (See Color Insert.)

including hair cells, may be more difficult to observe than in cells of the *Drosophila* wing. Also asymmetric localization may not occur in mammals. The initial analysis of the defects in *lp/lp* mice indicated a defect in the direction of the centrifugal movement of the developing kinocilium, suggesting that *vangl2* directs this movement, but does not influence the active movement itself. Considering that the developing kinocilium moves towards the distal part of each hair cell, localization of *vangl2* to the proximal membrane would suggest an inhibitory or repulsive role.

Some clues regarding the possible mechanism of action of *vangl2* come from a recent study in *Drosophila* (Lee et al., 2003). In this study, the authors demonstrate that a molecular complex, including *vang* and disc large (*dlg*), recruits vesicles to sites of new membrane formation (Fig. 6). Combining their results and the demonstrated genetic interaction of *vangl2* and *scrb1*, the authors hypothesize that *dlg*, lethal giant larva (*lgl*), *scrb* and *vang* may interact functionally with components of an undetermined vesicular transport apparatus to promote formation of plasma membranes with proper apical-basal polarity. The dynamic changes in localization of the PCP proteins are consistent with fast transport machinery that would be capable of mediating vesicle transport from the Golgi apparatus to the plasma membrane. It is not immediately clear how this type of transport could result in the generation of PCP, however for stereociliary bundles, it is possible to hypothesize that addition of membrane in the proximal region of a cell could generate a driving force that might cause the developing kinocilium to move towards the distal edge of the cell.

A second possible role for *vang*/*vangl2* has been suggested based on studies demonstrating physical interactions between *vang* and *dsh* (Park and Moon, 2002; Bastock et al., 2003), *dsh* and *pk* (Tree et al., 2002; Takeuchi et al., 2003), and *vang* and *pk* (Bastock et al., 2003). Since *dsh* ultimately becomes localized to the distal part of the cell, in contrast with *vang* and *pk* which become localized to the proximal side, these interactions presumably occur prior to the development of asymmetry. Recently, Tree et al. (2002) suggested that *vang, dsh*, and *pk* could form a complex that prevents the accumulation of *dsh* at the proximal side of wing epithelial cells. It should be noted, however, that while Tree et al. (2002) were able to show that overexpression of *pk* leads to a decrease in *dsh* translocation to the membrane, a second group (Bastock et al., 2003) was unable to corroborate this result.

Finally, Bellaïche et al. (2004) described the existence of a *vang*-dependent recruitment of Partner of Inscuteable (pins) at the anterior cortex of the sensory organ precursor of the *Drosophila* bristle lineage. In mammals, the pins homolog, Lethal Giant Larvae (*lgl*), has been shown to regulate microtubule stability (Du et al., 2002). This observation is interesting in light of the observation that *vangl2* regulates the movement of the kinocilium, a tubulin-rich structure that appears to dictate stereociliary bundle orientation (Montcouquiol et al., 2003).

6. Scribble 1

In 2000, Rachel et al. (2000) described the phenotype for a new spontaneous transgenic called Circletail (*crc*). *crc* heterozygotes have a curly tail while *crc* homozygotes dies at birth as a result of severe neural tube defects. Both of these phenotypes are similar to those observed in *lp* mice, suggesting that the *crc* mutation might be localized to a gene that acts in the same pathway as *vangl2*. In 2003 it was demonstrated that the *crc* phenotype is caused by a mutation in *scrb1* (a mammalian ortholog of Drosophila *scrb*) (Murdoch et al., 2003). *scrb1*/*scrb* codes for a cytoplasmic scaffolding protein that includes four PDZ domains while in *crc*, the protein is truncated such that the last two PDZ domains are lost (Murdoch et al., 2003). Based on the similar phenotypes between *crc* and *lp*, it seemed likely that the *crc* mutation would also lead to PCP defects in the ear. However, while the analysis of stereociliary bundle orientations in *crc* mice did indicate defects in PCP, these effects were considerably less severe than in *lp* mice and were limited to the second and third row outer hair cells (Montcouquiol et al., 2003). In contrast, PCP defects in the inner ears of animals that were double heterozygous for the *crc* and *lp* alleles were similar to the defects observed in *lp* homozygotes confirming a strong genetic interaction between *scrb1* and *vangl2*.

The demonstration that *scrb1* acts as a PCP gene in mammals was surprising in light of the role of *scrb* in *Drosophila*. *scrb* is classified as a cell, rather than planar cell, polarity gene and has been shown to be an early marker for the site of the future septate junction (analogous to the vertebrate tight junction) at the apical boundary of the basolateral compartment (Bilder and Perrimon, 2000; Fig. 6). *scrb* mutant larvae have loss of apical-basal polarity that leads to a loss of epithelial tissue structure. In these mutants, adherens junction proteins, including armadillo are mislocated, and apical transmembrane proteins exhibit both apical and basolateral distribution (Bilder et al., 2000). Two hypotheses have been suggested to explain the role of *scrb* in apical-basal polarity. *scrb* could restrict apical membrane determinants to the apical cell surface by assembling a physical "barrier" to separate apical and baso-lateral compartments. Alternatively, it may also play a role in polarized targeting of vesicles carrying apical proteins to the apical part of the cell. Finally, *scrb* has been shown to interact with other cell polarity genes including *dlg* (MAGUKs in mam-mals) and lethal giant larvae (*lgl*), and the localizations of these proteins show strong co-dependence (for review see Macara, 2004).

As discussed, *scrb1/scrb* are scaffolding proteins of the LAP (Leucine-rich repeat and PDZ) family (Bilder and Perrimon, 2000). They are large proteins that are composed of 16 leucine-rich repeats (LRR) in their N termini, followed by four PDZ (for post-synaptic density/Disc large/Z0-1) domains towards the C termini (Humbert et al., 2003). The LRR repeats are thought to be involved in modulating *ras*-family member signaling (Sternberg and Alberola-Ila, 1998), while PDZ domains are 90 amino acids stretches involved in protein–protein interactions with membrane or cytoskeletal proteins (Sheng and Sala, 2001).

A straightforward explanation for the phenotype in the *crc*/+, *lp*/+ double hetero-zygotes is that *scrb1* probably has a function very similar to *scrb* in *Drosophila*, and is necessary to establish the apical-basal polarity in inner ear hair cells, a necessary first step for the establishment of PCP. The lack of gross apical-basal disruptions in the *crc* homozygotes does not exclude the possibility that some apical proteins are mislocalized as a result of this mutation, and further experiments are clearly required to determine whether this is the case. It is also possible that *scrb1* plays a more direct role in the establishment of PCP in mammals, by specific targeting of some PCP proteins, or by forming a multiprotein complex that is necessary for anchoring and/or activating downstream cascades. The multiple PDZ domains within the *scrb1* protein are consistent with this hypothesis. In fact, we have recently confirmed that *vangl2* and *scrb1* physically interact, suggesting possible direct interactions *in vivo* (Montcouquiol M, Sans N, Wenthold R and Kelley MW, unpublished data). In the future, it will be interesting to determine whether other apical-basal polarity genes such as *lgl* or *dlg* are also involved in PCP in mammals.

7. Analogies with invertebrates

The remarkable degree of conservation of the PCP pathway at the genetic and molecular levels makes it tempting to speculate that an examination of an analogous system in *Drosophila* might lead to meaningful insights regarding the mechanisms that control stereociliary bundle orientation. In particular, the more straightforward asymmetric model of localization of *vang* in *Drosophila* wing cells might appear most applicable to the inner ear system; however, it may turn out that the *Drosophila* eye is the more appropriate analog. Fly wing cells are a homogenous population of identi-cal cell types, whereas PCP in the fly eye involves many different and specialized cell types, similar to the mammalian inner ear. With two different sensory hair cells and at least five supporting cell types, the cochlea is a complex system that can be favorably compared to the eight different photoreceptor cells that form each ommatidium. Also, Notch signaling, which has been shown to play a key role in the development of PCP and the determination of the R4 versus R3 cell fate in the eye, also regulates cell fate in the inner ear and seems to affect PCP, although the mechanisms for this effect have not been examined (Lanford et al., 1999).

8. Unique aspects of *Vangl2*

As mentioned before, the *vang* and *vangl2* genes code for putative four-pass transmembrane proteins with an intracellular N terminus domain making it unlikely that this protein acts as a receptor for an extracellular ligand. The C terminal portion of the protein contains a coiled-coil domain and a PDZ-binding domain. As discussed, there has been a high degree of conservation between *Drosophila vang* and mammalian *vangl2*; however, one of the more intriguing differences between the two proteins is the replacement of the cysteine at residue 464 in *vang* with a serine in *vangl2*. This substitution does not appear to create any obvious changes in the overall structure of the predicted protein. However, mutation of this serine to an asparagine leads to the *lp* phenotype, including PCP and neural tube defects (Kibar et al., 2001; Murdoch et al., 2001; Montcouquiol et al., 2003). More analysis will be required to understand the implication of this difference between mammals and the fly and its impact on the PCP cascade in both species.

9. Celsr1

Flamingo (*fmi*; also called *starry night*) is another important protein involved in the establishment of PCP in *Drosophila* (Chae et al., 1999; Lu et al., 1999; Usui et al., 1999). In fact, *fmi* was the first PCP protein to be shown to be asymmetrically localized in *Drosophila* wing (Usui et al., 1999). Prior to prehair outgrowth, and in a pattern similar to *vang*, *fmi* distribution becomes polarized along the proximal distal axis such that *fmi* is present at both the distal and proximal cell membranes but is excluded from the lateral sides of the cell. This distribution pattern is in contrast with *fz*, *vang*, *dsh*, and *pk*, all of which become restricted to either the distal or the proximal boundary of the wing cell (Usui, 1999; Feiguin et al., 2001). Asymmetric distribution of *fmi* is dependent on *fz* activity and, moreover, in the absence of *fmi*, there is disruption in the apical recruitment of *fz*, *vang*, dvl, and *pk* (Das et al., 2002; Rawls and Wolff, 2003). In the *Drosophila* eye, *fmi* is required in both cells of the R3 and R4 pair (Strutt et al., 2002).

In mammals, three *fmi* homologues have been described, *Celsr1-3* (for Cadherin EGF LAG Seven-pass G-type Receptor), with overlapping and distinct patterns of expression in embryonic and adult mouse (Hadjantonakis et al., 1998; Formstone and Little, 2002; Shima et al., 2002; Tissir et al., 2002). All three *Celsr* genes are expressed in the cochlea, although in different patterns, suggesting that each might serve a distinct function in the ear (Shima et al., 2002). In a recent study, two novel mouse mutants were described which displayed typical vestibular defects (head-shaking behavior) in heterozygotes and defects in formation of the neural tube, as well as embryonic death when homozygous for the mutation. Both mutants arose during the course of an ENU mutagenesis screen (Curtin et al., 2003), and were named *spin cycle* and *crash*. Subsequent positional cloning experiments demonstrated that both mutants carry independent missense mutations within the coding region of *Celsr1*, and both mutations affect the extracellular domain of the protein, within the cadherin-repeats (Curtin et al, 2003). Analysis of cochleae from both mutants indicated severe PCP defects, establishing *Celsr1* as the second molecule with a conserved function in PCP from flies to mammals. It is interesting that mutations in the extracellular domain lead to PCP defects, as it has been postulated for some time that the structure of the members of the *flamingo* family (7 pass transmembrane proteins) could involve some intracellular G-protein signaling. Clearly, the extracellular domain of *Celsr1* is necessary for its function, possibly through homophilic interactions (see Takeichi et al., 2000). If *Celsr1* mode of action in mammals is similar to that in *Drosophila*, we would expect to see its expression on both the proximal and distal side of the inner and outer hair cells, and can expect that many of the PCP molecules, including the *fz* receptors, *vangl2*, and Dvl would be mislocalized in the *crash* and *spin cycle* mutants. We therefore need to develop more tools, including antibodies, in order to determine the specific cellular localization of the three *Celsrs*, as well as, other PCP genes. Also, it would be interesting to determine if a null mutation for *Celsr1* would lead to a more dramatic disruption of PCP than in *crash* and *spin cycle*, both of which are probably hypomorphs. Finally, since there are three *Celsrs*, it would be of interest to see if mutations in either *Celsr2* and/or *Celsr 3* would lead to similar PCP defects, and also to determine how much redundancy exists between the three in mammals.

The *fmi* and *Celsr* genes encode atypical cadherins with large extracellular domains containing nine cadherin motifs, four EGF-like motifs, and two laminin G motifs (Hadjantonakis et al, 1998; Usui et al., 1999). In contrast to the classic-type cadherins, the carboxy-terminal intracellular tail of *fmi/Celsr* does not possess a catenin-binding sequence, nor does it bind to catenins. In addition, *fmi/Celsr* is also believed to be a seven-pass transmembrane (7TM) protein with similarity to members of the G-protein-coupled receptors. Whether *fmi/Celsr* is coupled to G proteins remains to be determined (Takeishi et al., 2000; Uemura and Shimada, 2003).

10. Ptk7

Using a gene-trapping screen, Lu et al. (2004) recently identified a new gene involved in PCP in the mammalian inner ear. This gene, called *Protein tyrosine kinase 7* (*PTK7*), encodes an evolutionary conserved transmembrane protein with a tyrosine kinase homology. The homozygous mutants for the gene die perinatally, and display similar phenotypes than those found in *lp/lp* animals including craniorachischisis. In the ear, bundle misorientation seems to be mainly restricted to the third row of outer hair cells, a phenotype similar to the *crc* mutants (Montcouquiol et al., 2003). The authors also found a weak genetic interaction between *vangl2* and *PTK7* when they generated a double hetrozygote *lp/+; PTK7/+*. The double heterozygotes exhibited spina bifida in most cases, but not the severe neural tube defects observed in the *lp/+; crc/+* animals, and PCP defects in the cochlea did not seem as severe either.

The PTK7 protein is predicted to contain an extracellular domain with seven immunoglobulin domains, a single transmembrane domain followed by an intracellular domain with tyrosine kinase homology. The extracellular immunoglobulin domain suggests that the protein could regulate PCP through adhesion, as has been suggested for the *Celsr1-3* molecules, and its cytoplasmic regions could also activate intracellular cascades. It is of interest to note that PTK7 seem to lack endogenous kinase activity.

11. Wnt-Frizzled signaling and stereociliary bundle reorientation

As described above, the *fz* receptor is one of the core PCP genes in *Drosophila*. The results of numerous studies have demonstrated that *fz* binds wingless (*wg*), a secreted glycoprotein that can act as both a short-range and long-range (up to 20 cell diameters) signaling molecule (Wodarz and Nusse, 1998). However, to date, a role for *wg* in PCP in *Drosophila* has not been demonstrated. In contrast, x*wnt*11, a vertebrate ortholog of *wg*, has been shown to play a role in convergence and extension during the formation of the *Xenopus* neural tube (Tada and Smith, 2000), a process that utilizes the same molecular pathway as PCP (reviewed in Mlodzik, 2002). These results suggest that orthologs of *wg* could play a role in PCP in vertebrates as well.

The family of vertebrate *wg* orthologs, referred to as *wnts*, includes 19 known genes in both mouse and human with their biological effects mediated through binding to different members of the *frizzled* family of receptors, of which there are ten members in mice and humans (www.stanford.edu/~rnusse/*wnt*window.html). *frizzled* receptors are characterized by a cysteine-rich ligand-binding domain where *wnts* bind, a seven transmembrane region, and a cytoplasmic terminal that in some *frizzled* receptors contain PDZ-binding domains (Hering and Sheng, 2002; Strutt, 2003). At present, *wnt* signaling has been shown to activate three distinct pathways: the *wnt*/β-catenin pathway, the *wnt*/Calcium pathway and the Jun-N-terminal kinase (JNK) pathway (Wodarz and Nusse, 1998; Kuhl et al., 2000; Huelsken and Birchmeier, 2001; Habas et al. 2003). While there is clear evidence that supports *wnt-fz* binding, and *wnt*-dependent activation of *fz*, the specificity of the ligand-receptor interactions remains unclear. This is especially true in vertebrates where there are large numbers of *wnt* and *fz* genes giving rise to the potential for a host of *wnt-fz* interactions and functional redundancies.

Within the developing mouse cochlea, the results of PCR and in situ screens have demonstrated that at least eight *wnts* and five *fzs* are expressed (Daudet et al., 2002; Dabdoub et al., 2003; Dabdoub and Kelley, unpublished data), suggesting a potential role for *wnt-fz* interactions in different aspects of inner ear development. In particular, the specific role of *wnt-fz* interactions in the development of uniform bundle orientation was examined recently using an *in vitro* approach to compensate for potential functional redundancy. Modulation of *wnt-fz* interactions resulted in misorientation of stereociliary bundles, however these effects were limited to outer hair cells (Table 1). Moreover, analysis of bundles at different developmental time points

Table 1
Modulation of Wnt-Frizzled signaling results in misorientation of outer hair cells. In order to perturb endogenous Wnt signaling, cochlear explants were treated with one of the following: 1) Medium conditioned with Wnt7A protein (Hall et al., 2000). 2) Secreted-frizzled-related-protein 1 (sFRP1; 50 μg/ml), known to bind to various Wnts and inhibit their activity (Rattner et al., 1997; Uren et al., 2000). 3) Medium conditioned with Wnt inhibitory factor 1 (WIF1), a secreted Wnt antagonist (Hsieh et al., 1999). 4) Sodium chlorate (30 μM), an inhibitor of the production of heparan sulfate proteoglycans that have been shown to play a key role in the extracellular distribution of Wnts (Kispert et al., 1996; Tsuda et al., 1999; Baeg et al., 2001; Dhoot et al., 2001). In contrast with controls, all of the treatments described above resulted in misorientation of stereociliary bundles in outer hair cells. Data are presented as average deviation from 0°, stereociliary bundle orientations were determined at the 5% and 12.5% positions along the basal-to-apical axis of the cochlea – determined as a percent of the total length. Control values represent pooled average deviations for all experiments

Position	Treatment				
	Control	Wnt7A-CM	sFRP1	WIF1-CM	NaClO$_3$
5%	12.1	37.1	25.6	52.7	27.6
12.5%	11.3	24.4	35.2	49.3	29.5

demonstrated that the effects of modulation of *wnt-fz* signaling on bundle orientation were limited to the reorientation step, rather than the initial centrifugal step, that was disrupted in *lp* mutants. These results clearly implicated a role for *wnt* signaling in the establishment of uniform bundle orientation and suggested that the role of *wg/wnt* in PCP may have expanded in vertebrates as compared with invertebrates.

One of the initial screens for *wnt* gene expression in the cochlea indicated that *wnt7a* might be the predominant *wnt* gene expressed in the developing sensory epithelium during the period of bundle orientation. Also, *in situ* hybridization revealed that expression of *wnt7a* is asymmetrically expressed in the cochlea, with strong expression in the pillar cells, at the time of stereociliary bundle reorientation. This distribution suggested that *wnt7a* emanating from the pillar cells could establish a gradient of *wnt7a* protein across the developing outer hair cells. However, there are two important caveats to this hypothesis. First, stereociliary bundle orientations are normal in *wnt7a* mutant mice, as well as in a number of other single gene mutations for *wnts* and *fz* (Dabdoub and Kelley, unpublished). Considering the overlap in spatio-temporal expression among different *wnt* and *fz* family members in vertebrates, the absence of PCP defects in single *wnt* or *fz* knockouts is not surprising considering the possibility of at least partial redundancy in function. This is certainly the case in *Drosophila* where studies have shown that *fz* and *fz2* are redundant *wg* receptors in embryonic development (Bhat, 1998; Kennerdell and Carthew, 1998; Bhanot et al., 1999; Chen and Struhl, 1999). Since this pattern of redundancy may also hold in mammals, or be exacerbated by the size of the *wnt* and *fz* gene families, it seems likely that it will be necessary to generate mice containing mutations in multiple *wnt* and/or *fz* genes before a clear PCP phenotype will occur.

A second caveat to a gradient hypothesis is the fact that the experimental design used for these experiments should have lead to the generation of a uniform concentration of *wnt-fz* signaling throughout the epithelium. It should also be considered that these data are consistent with a permissive role for *wnt* in which a specific range of *wnt-fz* signaling is required for reorientation but does not actually convey instructive information in terms of the direction of polarity within the epithelium. Clearly, further experiments are required to determine the specific role of *wnt* signaling within this system.

Despite the caveats discussed above, instructive gradients of biological morphogens represent one of the oldest hypotheses regarding the potential mechanisms for the generation of a uniform polarity within a plane of epithelial cells (reviewed in Teleman et al., 2001). If we apply this hypothesis to the developing cochlea, a potential source of instructive cues for the determination of cellular polarity would be a secreted factor that was produced by cells located along either the neural or abneural edge of the organ of Corti. Diffusion of this factor from a line source would result in the formation of a gradient leading to an asymmetric distribution of the soluble molecule across the plane of the hair cells (Fig. 8). Individual cells within the plane would be able to detect this gradient and to generate a uniform cellular polarity that would correspond with the direction of the gradient. There are several pieces of data that are consistent with this hypothesis. First, as discussed, the spatiotemporal gradient of stereociliary bundle polarization that exists between hair cells in different rows is consistent with increasing distance from a line source. The observation that inner hair cells and first row outer hair cells become polarized prior to second or third row outer hair cells suggests that the location of this source is nearer to the inner hair cells and first row outer hair cells. The second piece of data that supports

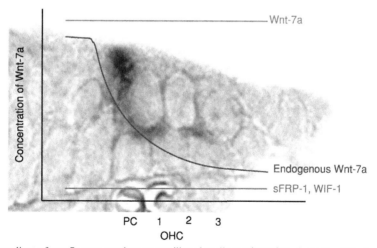

Fig. 8. A gradient of *wnt*-7a may regulate stereociliary bundle reorientation. A graph of the hypothetical gradient of the relative concentration of *wnt*-7a has been overlaid on a cross-section of the organ of Corti with expression of *wnt*-7a mRNA in the pillar cells illustrated by *in situ* hybridization (dark stain). If the pillar cells act as a line source of *wnt*-7a, then diffusion across the outer hair cell region would generate the illustrated concentration gradient. Addition or exogenous *wnt*-7a or treatment with *wnt*-inhibitors would alter the gradient as illustrated.

this hypothesis is the demonstration that expression of *wnt*7a is restricted to the developing pillar cells, located between the inner and outer hair cells, at the time of bundle reorientation. The localization of *wnt*7a in the pillar cells, between inner and outer hair cells, would suggest that inner and outer hair cells must respond differently to the putative orienting signal, since inner hair cell bundles are located on the proximal side of the cell relative to the pillar cells while outer hair cell bundles become localized on the distal side of each cell relative to the same putative source. In fact, treatment with *wnt*7a or *wnt* pathway inhibitors only affects outer hair cell reorientation. Therefore, these results suggest two intriguing hypotheses. First, that pillar cells may act as a line source for PCP signals within the organ of Corti, and second that the nature of those cues, or the responses of the hair cells in terms of bundle orientation may differ depending on the hair cell type.

12. Model

The existing data, although extremely limited, can be used to propose a preliminary model for the cellular and molecular interactions that occur during the development of uniform bundle orientation in the developing cochlea. The first sign of asymmetry within developing hair cells is the centrifugal migration of the nascent kinocilium from the center of the lumenal surface towards the distal end. Since the direction of this migration is non-random, it seems likely that *vang*l2, *scrb*1, *Celsr*1, and Ptk7 must all act prior to this migration in some way to guide the direction of movement. An analogy with the PCP in *Drosophila* would suggest that *Celsr*1 would become localized to both the proximal and distal boundaries of each hair cell, while *vang*l2 would become restricted to the proximal region of the hair cells. Continuing the *Drosophila* analogy, the restricted patterns of expression for *Celsr*1 and *vang*l2 would work in conjunction with a *fz*, or *fz*s, that would presumably become restricted to the distal end of each hair cell, to generate an asymmetric localization of *dsh* to the distal region of each cell. *dsh* would then act, possibly through small GTPases such as Rho or Rac, to stabilize or modulate some aspect of the actin or microtubule cytoskeleton. The resulting cytoskeletal changes might provide a scaffold that could be used to direct the movement of the kinocilium towards the distal edge of the cell. Following the centrifugal migration, a gradient of *wnt*7a protein emanating from the developing pillar cells would be used by each cell to refine the position of the developing bundle. In this case, the gradient of *wnt*7a would result in differential activation of *fz*, either across the entire cell, or, if *fz* becomes localized to the distal half of the cell as suggested above, just within the distal half of each cell. This gradient of differential activation would be used to slowly refine the position of the bundle, possibly through further modifications of the cytoskeleton. However, it is important to consider that this model is highly speculative and that there are a number of alternate scenarios that could also lead to uniform orientation.

One of the more intriguing aspects of this model is the idea that an organizing center may exist for regulating bundle orientation. In addition to the data presented in the section on *wnt-fz* interactions, there are several aspects of the normal and disrupted patterns of bundle orientation that are also consistent with this hypothesis.

First, there is a distinct gradient in the acquisition of appropriate bundle orientation that extends from inner hair cells across the three rows of outer hair cells. Second, distinct gradients are also observed for each of the known PCP mutants. Mutations in *scrb1, Celsr1*, and *Ptk7* are largely restricted to third row outer hair cells. Similarly, the defects in *vangl2* mutants, while not restricted to the third row, are far more severe in these cells than in the second or first row. These results are consistent with a multiply-redundant system in which defects in PCP signaling become more evident in those cells located farthest from the source of the polarizing signal as a result of progressive degradation.

The location of such an organizing center, as well as, the nature of the signals that might emanate from this center is unclear. One possibility would be the two rows of inner and outer pillar cells located between the inner hair cells and first row of outer hair cells. As discussed, these cells express *wnt7a* and also strongly express *vangl2* (Montcouquiol et al., 2003). However, as discussed, if the organizing center is located in the pillar cells, then the response of the inner hair cells to any polarizing signals must be fundamentally different from the response elicited from outer hair cells to the same signals. While this is possible, the existing data suggests that this is unlikely.

A second and perhaps more easily interpretable hypothesis would be that the organizing center is located adjacent to the neural edge of the row of inner hair cells. But, at this point there is no evidence to suggest that such a center exists in this location. Additional experiments are clearly required to address these different hypotheses.

13. Summary

The vertebrate inner ear represents one of the best examples of PCP. The generation of fields of uniformly oriented stereociliary bundles not only provides an easily assayed phenotype for PCP but the important functional implications of the generation of this polarization suggests a high degree of selective pressure to ensure appropriate development. Over the last two years our understanding of the molecular pathways that regulate the orientation of these bundles has increased dramatically. The results of both *in vivo* and *in vitro* studies have demonstrated that the functions of many of the genes that regulate PCP in *Drosophila* have been conserved in their mammalian orthologs. However, there are intriguing differences in both the collection of genes that are used to regulate polarization and in the cellular mechanisms and targets that are involved. Now that a number of key players have been identified, it seems likely that the next few years will include exciting insights into both the molecular pathways and cellular interactions that regulate PCP both in the ear and in other aspects of vertebrate and invertebrate biology as well.

References

Adler, P.N. 2002. Planar signaling and morphogenesis in *Drosophila*. Dev. Cell 2, 525–535.

Axelrod, J.D. 2001. Unipolar membrane association of *Disheveled* mediates frizzled planar cell polarity signaling. Genes. Dev. 15, 1182–1187.

Baeg, G.H., Lin, X., Khare, N., Baumgartner, S., Perrimon, N. 2001. Heparan sulfate proteoglycans are critical for the organization of the extracellular distribution of Wingless. Development 128, 87–94.

Barald, K.F., Kelley, M.W. 2004. From placode to polarization: New tunes in inner ear development. Development 131, 4119–4130.

Bastock, R., Strutt, H., Strutt, D. 2003. Strabismus is asymmetrically localized and binds to *prickle* and Disheveled during *Drosophila* planar polarity patterning. Development 130, 3007–3014.

Bellaïche, Y., Beaudoin-Massiani, O., Stüttem, I., Schweisguth, F. 2004. The planar cell polarity protein Strabismus promotes Pins anterior localization during asymmetric division of sensory organ precursor cells in *Drosophila*. Development 131, 469–478.

Bhanot, P., Fish, M., Jemison, J.A., Nusse, R., Nathans, J., Cadigan, K.M. 1999. Frizzled and Dfrizzled-2 function as redundant receptors for Wingless during Drosophila embryonic development. Development 126, 4175–4186.

Bhat, K.M. 1998. Frizzled and frizzled 2 play a partially redundant role in wingless signaling and have similar requirements to wingless in neurogenesis. Cell 23, 1027–1036.

Bilder, D., Perrimon, N. 2000. Localization of apical epithelial determinants by the basolateral PDZ protein Scribble. Nature 403, 676–680.

Bilder, D., Li, M., Perrimon, N. 2000. Cooperative regulation of cell polarity and growth by *Drosophila* tumor suppressors. Science 289, 113–116.

Chae, J., Kim, M.J., Goo, J.H., Collier, S., Gubb, D., Charlton, J., Adler, P.N., Park, W.J. 1999. The *Drosophila* tissue polarity gene *starry night* encodes a member of the protocadherin family. Development 126, 5421–5429.

Chen, C.M., Struhl, G. 1999. Wingless transduction by the frizzled and frizzled 2 proteins of *Drosophila*. Development 126, 5441–5452.

Corwin, J.T. 1981. Postembryonic production and aging in inner ear hair cells in sharks. J. Comp. Neurol. 201, 541–553.

Cotanche, D.A., Corwin, J.T. 1991. Stereociliary bundles reorient during hair cell development and regeneration in the chick cochlea. Hear. Res. 52, 379–402.

Curtin, J.A., Quint, E., Tsipouri, V., Arkell, R.M., Cattanach, B., Copp, A.J., Henderson, D.J., Spurr, N., Stanier, P., Fisher, E.M., Nolan, P.M., Steel, K.P., Brown, S.D., Gray, I.C., Murdoch, J.N. 2003. Mutation of *Celsr1* disrupts planar polarity of inner ear hair cells and causes severe neural tube defects in the mouse. Curr. Biol. 13, 1129–1133.

Dabdoub, A., Donohue, M.J., Brennan, A., Wolf, V., Montcouquiol, M., Sassoon, D.A., Rubin, J.S., Salinas, P.C., Kelley, M.W. 2003. *wnt* signaling mediates reorientation of outer hair cell stereociliary bundles in the mammalian cochlea. Development 130, 2375–2384.

Darken, R.S., Scola, A.M., Rakeman, A.S., Das, G., Mlodzik, M., Wilson, P.A. 2002. The planar polarity gene strabismus regulates convergent extension movements in *Xenopus*. EMBO J. 21, 976–985.

Das, G., Reynolds-Kenneally, J., Mlodzik, M. 2002. The atypical cadherin Flamingo links Frizzled and Notch signaling in planar polarity establishment in the *Drosophila* eye. Dev. Cell 2, 655–666.

Daudet, N., Ripoll, C., Moles, J.P., Rebillard, G. 2002. Expression of Wnt and Frizzled gene families in the postnatal rat cochlea. Brain Res. Mol. Brain Res. 105, 98–107.

Denman-Johnson, K., Forge, A. 1999. Establishment of hair bundle polarity and orientation in the developing vestibular system of the mouse. J. Neurocytol. 28, 821–835.

Dhoot, G.K., Gustafsson, M.K., Ai, X., Sun, W., Standiford, D.M., Emerson, C.P. Jr. 2001. Regulation of Wnt signaling and embryo patterning by an extracellular sulfatase. Science 293, 1663–1666.

Du, Q., Taylor, L., Compton, D.A., Macara, I.G. 2002. LGN blocks the ability of NuMA to bind and stabilize microtubules. A mechanism for mitotic spindle assembly regulation. Curr. Biol. 12, 1928–1933.

Eaton, S. 2003. Cell biology of planar polarity transmission in the *Drosophila* wing. Mech. Dev. 120, 1257–1264.

Feiguin, F., Hannus, M., Mlodzik, M., Eaton, S. 2001. The ankyrin-repeat protein Diego mediates Frizzled-dependent planar polarization. Dev. Cell 1, 93–101.

Formstone, C.J., Little, P.F. 2001. The flamingo-related mouse *Celsr* family (*Celsr1-3*) genes exhibit distinct patterns of expression during embryonic development. Mech. Dev. 109, 91–94.

Goto, T., Keller, R. 2002. The planar cell polarity gene strabismus regulates convergence and extension and neural fold closure in *Xenopus*. Dev. Biol. 247, 165–181.

Habas, R., Dawid, I.B., He, X. 2003. Coactivation of Rac and Rho by Wnt/Frizzled signaling is required for vertebrate gastrulation. Genes Dev. 15, 295–309.

Hackney, C.M., Furness, D.N. 1995. Mechanotransduction in vertebrate hair cells: Structure and function of the stereociliary bundle. Am. J. Physiol. 268, C1–C13.

Hadjantonakis, A.K., Formstone, C.J., Little, P.F.R. 1988. mCelsr1 is an evolutionarily conserved seven-pass transmembrane receptor and is expressed during mouse embryonic development. Mech. Dev. 78, 91–95.

Hall, A.C., Lucas, F.R., Salinas, P.C. 2000. Axonal remodeling and synaptic differentiation in the cerebellum is regulated by WNT-7a signaling. Cell 100, 525–535.

Hering, H., Sheng, M. 2002. Direct interaction of Frizzled-1, -2, -4, and -7 with PDZ domains of PSD-95. FEBS Lett. 19, 185–189.

Hsieh, C.-J., Kodjabachian, L., Rebbert, M.L., Rattner, A., Smallwood, P.M., Samos, C.H., Nusse, R., Dawid, I.B., Nathans, J. 1999. A new secreted protein that binds to Wnt proteins and inhibits their activities. Nature 398, 431–436.

Huelsken, J., Birchmeier, W. 2001. New aspects of Wnt signaling pathways in higher vertebrates. Curr. Opin. Gene. Devel. 11, 547–553.

Humbert, P., Russell, S., Richardson, H. 2003. Dlg, Scribble and Lgl in cell polarity, cell proliferation and cancer. Bioessays 25, 542–553.

Jones, J.E., Corwin, J.T. 1993. Replacement of lateral line sensory organs during tail regeneration in salamanders: Identification of progenitor cells and analysis of leukocyte activity. J. Neurosci. 13, 1022–1034.

Kennerdell, J.R., Carthew, R.W. 1998. Use of dsRNA-mediated genetic interference to demonstrate that frizzled and frizzled 2 act in the wingless pathway. Cell 23, 1017–1026.

Kibar, Z., Vogan, K.J., Groulx, N., Justice, M.J., Underhill, D.A., Gros, P. 2001. *Ltap*, a mammalian homolog of *Drosophila Strabismus/Van Gogh*, is altered in the mouse neural tube mutant Loop-tail. Nat. Genet. 28, 251–255.

Kispert, A., Vainio, S., Shen, L., Rowitch, D.H., McMahon, A.P. 1996. Proteoglycans are required for maintenance of Wnt-11 expression in the ureter tips. Development 122, 3627–3637.

Kuhl, M., Sheldahl, L.C., Park, M., Miller, J.R., Moon, R.T. 2000. The Wnt/Ca^{2+} pathway: A new vertebrate Wnt signaling pathway takes shape. Trends Genet. 16, 279–283.

Lanford, P.J., Lan, Y., Jiang, R., Lindsell, C., Weinmaster, G., Gridley, T., Kelley, M.W. 1999. Notch signalling pathway mediates hair cell development in mammalian cochlea. Nat. Genet. 21, 289–292.

Lee, O.K., Frese, K.K., James, J.S., Chadda, D., Chen, Z.H., Javier, R.T., Cho, K.O. 2003. Discs-Large and Strabismus are functionally linked to plasma membrane formation. Nat. Cell Biol. 5, 987–993.

Lu, B., Usui, T., Uemura, T., Jan, L., Jan, Y.N. 1999. Flamingo controls the planar polarity of sensory bristles and asymmetric division of sensory organ precursors in *Drosophila*. Curr. Biol. 9, 1247–1250.

Lu, X., Borchers, A.G., Jolicoeur, C., Rayburn, H., Baker, J.C., Tessier-Lavigne, M. 2004. PTK7/CCK-4 is a novel regulator of planar cell polarity in vertebrates. Nature 430, 93–98.

Macara, I.G. 2004. Parsing the polarity code. Nat. Rev. Mol. Cell. Biol. 5, 220–231.

Mlodzik, M. 2002. Planar cell polarization: Do the same mechanisms regulate Drosophila tissue polarity and vertebrate gastrulation? Trends Genet. 18, 564–571.

Montcouquiol, M., Rachel, R.A., Lanford, P.J., Copeland, N.G., Jenkins, N.A., Kelley, M.W. 2003. Identification of *vangl2* and *scrb* 1 as planar polarity genes in mammals. Nature 423, 173–177.

Murdoch, J.N., Doudney, K., Paternotte, C., Copp, A.J., Stanier, P. 2001. Severe neural tube defects in the loop-tail mouse result from mutation of Lpp1, a novel gene involved in floor plate specification. Hum. Mol. Genet. 10, 2593–2601.

Murdoch, J.N., Henderson, D.J., Doudney, K., Gaston-Massuet, C., Phillips, H.M., Paternotte, C., Arkell, R., Stanier, P., Copp, A.J. 2003. Disruption of scribble (*scrb* 1) causes severe neural tube defects in the Circletail mouse. Hum. Mol. Genet. 12, 87–98.

Park, M., Moon, R.T. 2002. The planar cell-polarity gene *stbm* regulates cell behaviour and cell fate in vertebrate embryos. Nat. Cell Biol. 4, 20–25.

Rachel, R.A., Murdoch, J.N., Beermann, F., Copp, A.J., Mason, C.A. 2000. Retinal axon misrouting at the optic chiasm in mice with neural tube closure defects. Genesis 27, 32–47.

Rattner, A., Hsieh, J.C., Smallwood, P.M., Gilbert, D.J., Copeland, N.G., Jenkins, N.A., Nathans, J. 1997. A family of secreted proteins contains homology to the cysteine-rich ligand-binding domain of frizzled receptors. Proc. Natl. Acad. Sci. 94, 2859–2863.

Rawls, A.S., Wolff, T. 2003. Strabismus requires Flamingo and Prickle function to regulate tissue polarity in the *Drosophila* eye. Development 130, 1877–1887.

Sheng, M., Sala, C. 2001. PDZ domains and the organization of supramolecular complexes. Annu. Rev. Neurosci. 24, 1–29.

Sher, A.E. 1971. The embryonic and postnatal development of the inner ear of the mouse. Acta Otolaryngol. Suppl. 285, 1–77.

Shima, Y., Copeland, N.G., Gilbert, D.J., Jenkins, N.A., Chisaka, O., Takeichi, M., Uemura, T. 2002. Differential expression of the seven-pass transmembrane cadherin genes Celsr1-3 and distribution of the Celsr2 protein during mouse development. Dev. Dyn. 223, 321–332.

Sternberg, P.W., Alberola-Ila, J. 1998. Conspiracy theory: RAS and RAF do not act alone. Cell 95, 447–450.

Strutt, D. 2003. Frizzled signaling and cell polarization in *Drosophila* and vertebrates. Development 130, 4501–4513.

Strutt, D., Johnson, R., Cooper, K., Bray, S. 2002. Asymmetric localisation of Frizzled and the determination of Notch-dependent cell fate in the *Drosophila* eye. Curr. Biol. 12, 813–824.

Strutt, D. 2001. Asymetric localization of Frizzled and the establishment of cell polarity in the *Drosophila* wing. Mol. Cell 7, 367–375.

Tada, M., Smith, J.C. 2000. Xwnt11 is a target of Xenopus Brachyury: Regulation of gastrulation movements via Disheveled, but not through the canonical Wnt pathway. Development 127, 2227–2238.

Takeichi, M., Nakagawa, S., Aono, S., Usui, T., Uemura, T. 2000. Patterning of cell assemblies regulated by adhesion receptors of the cadherin superfamily. Philos. Trans. R. Soc. Lond. B. Biol. Sci. 355, 885–890.

Takeuchi, M., Nakabayashi, J., Sakaguchi, T., Yamamoto, T.S., Takahashi, H., Takeda, H., Ueno, N. 2003. The prickle-related gene in vertebrates is essential for gastrulation cell movements. Curr. Biol. 13, 674–679.

Taylor, J., Abramova, N., Charlton, J., Adler, P.N. 1998. *Van Gogh*: A new *Drosophila* tissue polarity gene. Genetics 150, 199–210.

Teleman, A.A., Strigini, M., Cohen, S.M. 2001. Shaping morphogen gradients. Cell 105, 559–562.

Tissir, F., De-Backer, O., Goffinet, A.M., Lambert de Rouvroit, C. 2002. Developmental expression profiles of Celsr (Flamingo) genes in the mouse. Mech. Dev. 112, 157–160.

Tree, D.R.P., Shulman, J.M., Rousset, R., Scott, M.P., Gubb, D., Axelrod, J.D. 2002. Prickle mediates feedback amplification to generate asymmetric planar cell polarity signaling. Cell 109, 371–381.

Tsuda, M., Kamimura, K., Nakato, H., Archer, M., Staatz, W., Fox, B., Humphrey, M., Olson, S., Futch, T., Kaluza, V., Siegfried, E., Stam, L., Selleck, S.B. 1999. The cell-surface proteoglycan Dally regulates Wingless signalling in *Drosophila*. Nature 400, 276–280.

Uemura, T., Shimada, Y. 2003. Breaking cellular symmetry along planar axes in *Drosophila* and vertebrates. J. Biochem. 134, 625–630.

Uren, A., Reichsman, F., Anest, V., Taylor, W.G., Muraiso, K., Bottaro, D.P., Cumberledge, S., Rubin, J. S. 2000. Secreted frizzled-related protein-1 binds directly to Wingless and is a biphasic modulator of Wnt signaling. J. Biol. Chem. 275, 4374–4382.

Usui, T., Shima, Y., Shimada, Y., Hirano, S., Burgess, R.W., Schwarz, T.L., Takeichi, M., Uemura, T. 1999. Flamingo, a seven-pass transmembrane cadherin, regulates planar cell polarity under the control of Frizzled. Cell 98, 585–595.

Veeman, M.T., Axelrod, J.D., Moon, R.T. 2003. A second canon: Functions and mechanisms of b-catenin-independent Wnt signaling. Dev. Cell 5, 367–377.

Wodarz, A., Nusse, R. 1998. Mechanisms of Wnt signaling in development. Annu. Rev. Cell. Dev. Biol. 14, 59–88.

Wolff, T., Rubin, G. 1998. strabismus, a novel gene that regulates tissue polarity and cell fate decision in *Drosophila*. Development 125, 1149–1159.

Woolley, S.M., Wissman, A.M., Rubel, E.W. 2001. Hair cell regeneration and recovery of auditory thresholds following aminoglycoside ototoxicity in Bengalese finches. Hear. Res. 153, 181–195.

Yang, C.-H., Axelrod, J.D., Simon, M.A. 2002. Regulation of Frizzled by Fat-like cadherins during planar polarity signalling in the *Drosophila* compound eye. Cell 108, 675–688.

Yoshida, N., Liberman, M.C. 1999. Stereociliary anomaly in the guinea pig: Effects of hair bundle rotation on cochlear sensitivity. Hear Res. 131, 29–38.

Morphogenetic cell movements shaping the zebrafish gastrula

Jason R. Jessen and Lilianna Solnica-Krezel

Department of Biological Sciences, Vanderbilt University, Nashville, Tennessee 37235

Contents

1. Introduction

A diverse array of cell movement behaviors contributes to the development of multicellular organisms and the maintenance of normal adult physiology. Deregulated cell migration occurs in a variety of human diseases including cancer, where it promotes tissue invasion and tumor metastasis, the spread of cells from the primary

Advances in Developmental Biology
Volume 14 ISSN 1574-3349
DOI: 10.1016/S1574-3349(04)14007-6

tumor to distant tissues. Our ability to develop therapies for such devastating diseases will require a comprehensive and integrated understanding of cell movement behaviors *in vivo*, at both the cellular and molecular levels. Currently, we understand cell migration to be a multistep process that entails modulation of adhesion, dynamic interactions between the moving cell and its environment, changes in cell shape, polarity, membrane protrusive activity, and redistribution of intracellular and membrane proteins. A common theme to all types of movements is a cell's ability to modify its internal architecture in response to stimuli. Whether it is *Dictyostelium discoideum* amoebae chemotaxing up a cAMP gradient, leukocytes transmigrating across an endothelium, or zebrafish mesodermal cells converging towards the dorsal embryonic axis, cell migration and its associated movement behaviors involves specific changes in the actin cytoskeleton and microtubule network.

Cell movement has been studied in a variety of *in vitro* and *in vivo* systems, in the context of individual cells, groups of cells, and epithelial sheets. Recent years have witnessed a resurgence in the number of studies on cell movement during vertebrate gastrulation. During embryogenesis, gastrulation is an important phase characterized by cell fate specification and massive tissue rearrangements driven by widespread and complex cell movement behaviors. Much of our knowledge about gastrulation cell movements comes from studies of the frog (*Xenopus laevis*), the killifish (*Fundulus heteroclitus*), and the zebrafish (*Danio rerio*). In the picture emerging from these studies, the complexity of cell movement behaviors operating before, during, and after gastrulation is regulated by an intricate and diverse set of signaling networks. However, with complexity comes excitement, and it is indeed an exciting time to study the cellular and molecular mechanisms underlying cell movements during gastrulation.

This chapter reviews our current knowledge of the morphogenetic processes shaping the early zebrafish embryo. We begin with an overview of gastrulation cell movements followed by a detailed discussion of convergence and extension and its associated cell movement behaviors. Then we examine the molecules and signaling pathways that orchestrate and mediate these behaviors. For additional recent reviews on zebrafish gastrulation cell movements see Heisenberg and Tada (2002), Myers et al. (2002a), and Solnica-Krezel and Cooper (2002).

2. Overview of early morphogenetic processes in zebrafish

During early embryogenesis, inductive and morphogenetic processes cooperate to ensure that the three germ layers—ectoderm, mesoderm, and endoderm—are specified, and correctly positioned and shaped within the developing embryo. As a result, an initially simple blastula consisting of many blastomeres piled atop a large yolk cell, is transformed into an embryo with defined anteroposterior and dorsoventral axes. Three main morphogenetic processes operate concurrently during gastrulation to shape the early embryo including epiboly, internalization, and convergence and extension. While epiboly and internalization are completed by the end of the gastrula period, at 10 hours post-fertilization (hpf), convergence and extension movements continue through segmentation. For a detailed description and images of zebrafish embryonic development, see zfin.org/zf_info/zfbook/stages/stages.html and Kimmel et al. (1995). While reading this chapter, it is important to recognize that these morphogenetic processes are very dynamic, and overlap in time and space, such that a given cell, depending on its position, can simultaneously engage in more than one process. Therefore, the movement behaviors employed by a cell to achieve its appropriate position within the developing embryo are quite diverse, both in their execution and molecular regulation (see Section 4).

2.1. Epiboly

Epiboly is the first morphogenetic process to shape the developing embryo. At the beginning of epiboly (4.3 hpf), the embryo consists of three main tissue types, an inner large multinucleated yolk cell (yolk syncytial layer), an outer epithelial layer that envelopes the embryo (enveloping layer), and the blastoderm deep cells located in between (Fig. 1). Epiboly movements thin and spread the blastoderm vegetally and continue until 10 hpf when the yolk is completely covered (Fig. 1). The process of epiboly entails both interactions *between*, as well as mechanisms intrinsic *to*, each of these tissues. For example, the yolk cell contains a dynamic microtubule network that is required for epiboly movements of nuclei located within the yolk syncytial layer (Strähle and Jesuthasan, 1993; Solnica-Krezel and Driever, 1994). Moreover, epiboly of the syncytium continues upon removal of all blastomeres, consistent with the notion that movement of this tissue is autonomously driven and independent of the blastoderm (Trinkaus, 1951). In contrast, the enveloping layer covering the embryo may play a more static role during epiboly, being "pulled" vegetally by the yolk syncytial layer to which it is tightly linked at its margin (Betchaku and Trinkaus, 1978). Determining the role of the deep cells of the blastoderm during epiboly has been more problematic. In the *half baked* (*hab*) and *volcano* (*vol*) mutants, epibolic movements of both the yolk syncytial layer and enveloping layer occur normally while those of the deep cells are arrested by 60–70% epiboly (Kane et al., 1996; Solnica-Krezel et al., 1996). This suggests a movement behavior intrinsic to the deep

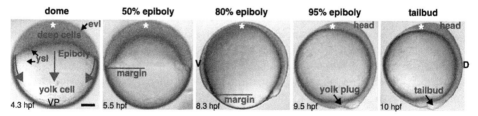

Fig. 1. The morphogenetic movement of epiboly thins and spreads the blastoderm and yolk syncytial layer over the yolk towards the vegetal pole. In the images of 80% epiboly to tailbud stage embryos, ventral (V) is to the left and dorsal (D) is to the right. evl, enveloping layer; hpf, hours post-fertilization; VP, vegetal pole; ysl, yolk syncytial layer (where ysl nuclei are located). In each image, the asterisk denotes the animal pole. Scale bar represents 100 μm.

cells is necessary for deep cells to spread to the vegetal pole. One such movement is radial intercalation, whereby deeper cells of the blastoderm move between superficial neighbors, and thus spread and thin the tissue (Kimmel and Warga, 1987; Wilson et al., 1995). It was recently shown that the *hab* locus encodes E-cadherin (D. Kane, University of Rochester, personal communication), a cell adhesion molecule that mediates homophilic protein–protein interactions and is typically associated with formation and maintenance of epithelial structures (reviewed in Gumbiner, 2000 and Yagi and Takeichi, 2000). This is a very exciting discovery in part because of recent findings showing that cadherin-based adhesion can be regulated by integrin interactions with the extracellular matrix (Marsden and Desimone, 2003; Sakai et al., 2003). Furthermore, current data also implicates cadherins in adhesion-activated intracellular signaling pathways that influence cytoskeletal organization (reviewed in Yap and Kovacs, 2003). It will now be important to identify the molecular nature of E-cadherin-based adhesion and/or signaling in the deep cells, and to determine its contribution to epiboly movements. Epiboly also plays a role in the overall antero-posterior elongation of embryonic tissues, as the vegetal spreading of tissues during epiboly leads to a longer anteroposterior axis. Intriguingly, the axial mesen-doderm is also broadened in *hab* (Kane et al., 1996) and *vol* (Solnica-Krezel et al., 1996) epiboly mutants. Defective convergence movements in these mutants may be secondary to the reduced extension/epiboly of embryonic tissues. However, it is possible that Hab/E-cadherin may influence convergence movements, independent of its role in epiboly.

2.2. Internalization

The second major morphogenetic process to shape the embryo is internalization, whereby prospective mesodermal and endodermal cells at the blastoderm margin move beneath the superficial future ectodermal cells (Fig. 2A). It remains a matter of debate (however, see below) whether internalization occurs via involution (the move-ment of a sheet of cells around the edge of the margin) or by ingression (the burrowing of individual cells from the superficial epiblast into the deep hypoblast) (Trinkaus, 1984; Warga and Kimmel, 1990; Trinkaus, 1996; Kane and Adams, 2002). The process of internalization marks the beginning of the gastrula period (6 hpf/ ~55% epiboly). This time point is characterized by a pause in epiboly and formation of the germ ring, a piling of internalized cells at the margin, immediately followed by formation of the embryonic shield, a thickening of the blastoderm margin on the future dorsal side of the embryo (Fig. 2A and B). The embryonic shield represents a large portion of the zebrafish equivalent to the Spemann-Mangold organizer, which, as in other vertebrates, plays an important role in both patterning the embryo and controlling gastrulation movements (reviewed in Hibi et al., 2002). Blastoderm cell internalization takes place around the entire circumference of the margin, but may begin earlier on the dorsal side, and continues throughout epiboly (reviewed in Kane and Adams, 2002). As epiboly and internalization movements proceed, two tissue layers (epiblast and hypoblast) become visible (Fig. 2C). Clues into the molecular control of internalization came from analysis of maternal-zygotic *one-eyed pinhead*

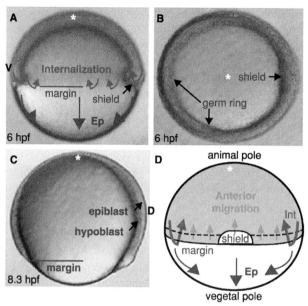

Fig. 2. The morphogenetic movement of mesendoderm internalization marks the beginning of gastrulation at approximately 6 hpf, and is followed shortly thereafter by the anterior migration of prospective mesendodermal cells towards the animal pole. (A) Internalization of marginal cells occurs around the circumference of the embryo (arrows). (B) Internalization movements are particularly noticeable on the dorsal side, as evidenced by a dorsal thickening called the shield. (C) Internalization of superficial cells forms two tissue layers, the inner hypoblast (giving rise to mesoderm and endoderm) underlying the outer epiblast (giving rise to neural and non-neural ectoderm). (D) Anterior migration of internalized cells occurs most prominently along the dorsal embryonic axis, and in conjunction with epiboly movements, contributes to anteroposterior elongation of the embryo. Ep, epiboly; Int, internalization. In each image, the asterisk denotes the animal pole. In A–C, ventral (V) is to the left and dorsal (D) is to the right.

(mz*oep*) mutant embryos, which are defective in Nodal signaling and whose blastoderm margin cells fail to internalize (Carmany-Rampey and Schier, 2001). This study demonstrated that single wild-type blastoderm margin cells transplanted into the margin of a mz*oep* host are capable of autonomous internalization and subsequent mesendoderm formation. The fact that internalization of these single cells does not require interactions between groups of cells is consistent with the notion that ingression may be the predominant mode of cell internalization (Carmany-Rampey and Schier, 2001; see also Heisenberg and Tada, 2002). For a comprehensive review of epiboly and internalization cell movements in zebrafish, see Kane and Adams (2002).

2.3. Anterior migration

Upon internalization, nascent mesendodermal progenitor cells migrate anteriorly towards the animal pole while later internalizing cells continue their vegetally directed epiboly movements (Fig. 2D; Warga and Kimmel, 1990). Therefore, similar to epiboly, internalization followed by anterior migration of hypoblast cells also contributes to anteroposterior extension of the embryonic axis. What are the cellular and molecular mechanisms controlling the anterior movement of dorsal mesendoderm? Do these cells exhibit active and directed anterior migration or do newly internalized cells passively push them away from the margin towards the animal pole? It has been suggested that internalized paraxial cells migrate as loosely associated groups while axial cells exhibit an epithelial morphology and only cells at the anterior edge actively migrate (Heisenberg and Tada, 2002). Furthermore, our lab has observed that lateral and ventral mesodermal cells migrate largely as individuals (D.S.S. and L.S.-K., unpublished data). Three recent studies on Wnt11 (Ulrich et al., 2003), Phosphoinositide 3-kinase (Montero et al., 2003), and Stat3 (Yamashita et al., 2002) shed some light on the molecular control of anterior migration and will be discussed in Section 4.

2.4. Convergence and extension

In addition to epiboly and internalization, convergence and extension cell movements are the major morphogenetic force shaping tissues in the zebrafish gastrula. While the majority of convergence and extension movements occur during gastrula stages (6 through 10 hpf), they also continue into segmentation stages. It is during the late gastrula and early segmentation stages when the initially round embryo becomes ovoid with a clear dorsal axis, anterior head rudiment, and posterior tailbud. The process of convergence and extension encompasses several cell movement behaviors that contribute to mediolateral narrowing (convergence) and anteroposterior lengthening (extension) of the embryonic axis (Fig. 3A,a and B,b). The consequence of defective convergence and extension movements is exemplified in *trilobite* mutant embryos (Fig. 3C,c; and see below). Gastrulation defects are apparent by morphology as well as by using whole-mount *in situ* hybridization to monitor gene expression in key tissues undergoing morphogenetic movements (Fig. 3D).

In fish, endodermal cells engaged in convergence and extension move solely as individuals (Warga and Kimmel, 1990; Warga and Nüsslein-Volhard, 1999), mesodermal cells move as individuals or in small groups, while ectodermal cells move as an epithelial-like sheet (Warga and Kimmel, 1990; Trinkaus et al., 1991; Trinkaus et al., 1992; Concha and Adams, 1998). By contrast, both mesodermal and ectodermal cells in *Xenopus* gastrulae move almost exclusively as stiff, coherent sheets and thereby exert an active influence on the behaviors of surrounding tissues (Keller et al., 2000). This observation may partly explain why in fish gastrulae, the three main morphogenetic movements occur largely independent from one another. For example, it was shown that zebrafish mutant embryos with severe defects in convergence and extension cell movements are capable of successfully completing epiboly and internalization (Ho and Kane, 1990; Solnica-Krezel et al., 1996; Marlow et al., 1998). The types of cell behaviors underlying convergence and extension movements in different regions of the embryo are subject to a variety of molecular and mechanical influences. Below we discuss these cell behaviors and how they control the speed and magnitude of convergence and extension movements within different gastrula domains.

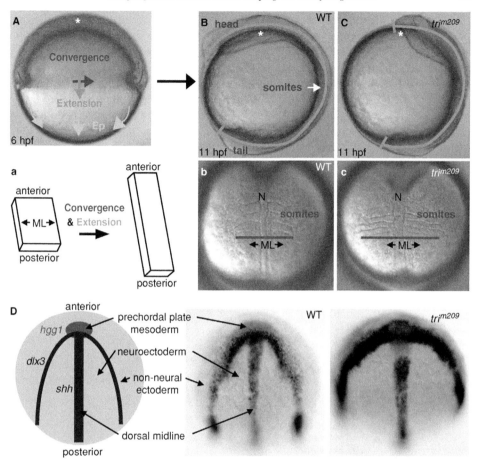

Fig. 3. (A,a) Convergence and extension movements contribute significantly to shaping the early embryo by narrowing tissues mediolaterally while simultaneously elongating them from head to tail. (B,b and C,c) In *tri* mutant embryos, impairment of convergence and extension cell movements result in an embryo that has a shortened anteroposterior axis (green lines) and is broadened mediolaterally (ML). (D) A typical whole-mount *in situ* hybridization analysis demonstrating how the use of combinations of gene expression probes that mark specific embryonic tissues can suggest disrupted convergence and extension movements. Note that the homozygous *tri* mutant embryo exhibits posteriorly shifted prechordal plate mesoderm (here, labeled in red with *hgg1*) and axial tissue (here, labeled with *shh*), and a mediolaterally broadened neural plate (here, the border between neural and non-neural ectoderm is labeled with *dlx3*). While *in situ* analyses are informative, they do not reveal anything about the underlying cellular defects. The asterisk denotes the animal pole. (See Color Insert.)

3. Domains of convergence and extension movements

An important implication for the types of cell behaviors employed during gastrulation is the fact that each germ layer participates in convergence and extension movements. It was the pioneering work of J. P. Trinkaus that first described how convergence and extension movements varied depending on a cell's position, along dorsoventral and anteroposterior axes, within the *Fundulus* gastrula (Trinkaus et al., 1992; Trinkaus, 1998). Recent work has revealed a similar diversity of cell movements in zebrafish. These studies defined three distinct domains of convergence and extension along the dorsoventral embryonic axis of the gastrula (Sepich et al., 2000; Myers et al., 2002b): a dorsal domain of strong extension with limited convergence movements (Fig. 4A); a lateral domain of increasing convergence and extension movements (Fig. 4B); and a ventral domain lacking convergence and extension movements (Fig. 4C). It has been proposed that these domains are specified by the ventral to dorsal gradient of bone morphogenetic protein (Bmp) activity (Myers et al., 2002b; reviewed in Myers et al., 2002a). Currently it is not known whether there is a gradual or sharp transition between these domains. However, as discussed below, distinct cell movement behaviors underlie the different types of individual and population cell movements that occur within these convergence and extension domains.

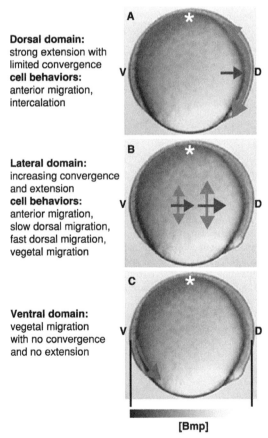

Fig. 4. Domains of convergence and extension in the zebrafish gastrula. Ventral (V) is to the left and dorsal (D) is to the right. The ventral to dorsal Bmp activity gradient is schematized at the bottom. The asterisk denotes the animal pole.

3.1. Dorsal domain

In general, the dorsal domain of convergence and extension movements is characterized by limited convergence and strong extension of the embryonic axis. It is important to note that the rate of convergence movements in the dorsal domain is influenced by cellular location (i.e., paraxial or axial tissues). Analyses of cell movements have shown that the speed of converging dorsolateral mesendodermal and ectodermal cell populations slows as these cells approach the dorsal axis or midline (Kimmel et al., 1994; Heisenberg et al., 2000; Sepich et al., 2000). Current data suggest that besides anterior migration and epiboly, mediolateral intercalation may be the major cell behavior driving strong extension in the dorsal domain (Warga and Kimmel, 1990; Kimmel et al., 1994; Concha and Adams, 1998; Heisenberg et al., 2000; Topczewski et al., 2001; Myers et al., 2002b; Glickman et al., 2003). This cell behavior is best described in *Xenopus* gastrulae, where studies on tissue explants have shown that dorsal mesoderm simultaneously converges and extends (convergent extension) via large-scale cellular rearrangements (Shih and Keller, 1992a,b; Keller et al., 2000). In the mediolateral intercalation behavior model (Shih and Keller, 1992a; Keller et al., 2000), cells are initially nonpolarized exhibiting rounder morphologies and randomly directed membrane protrusions. Then, presumably after receiving the proper physical and/or molecular signals, protrusions are formed in both the lateral and medial directions where they attach and exert traction on adjacent cell surfaces. These bipolar cells elongate with their long axes oriented perpendicular to the dorsal midline and intercalate between one another, thereby narrowing and elongating the embryonic axis. In zebrafish gastrulae, mediolateral intercalation behavior, as measured by the number of neighbor exchanges, is displayed by dorsal mesodermal cells (Glickman et al., 2003). However, mediolateral intercalations underlying simultaneous convergence and extension (convergent extension) in zebrafish may be limited to notochord-forming axial mesoderm (Fig. 5A). In paraxial somitic mesoderm, the rate of dorsal convergence does not equal the rate of extension; instead, mediolateral intercalations may be contributing to dorsoventral tissue thickening and not anteroposterior tissue extension (Glickman et al., 2003). Intriguingly, in *no tail/brachyury* mutant embryos with severely disrupted convergence movements, dorsal mesoderm is able to partially extend (Glickman et al., 2003). Therefore, at this stage and in this region, convergence and extension are not necessarily linked and other cell behaviors must be involved, such as radial intercalation (epiboly) and anterior migration.

Reminiscent of *Xenopus* cells engaged in mediolateral intercalation, zebrafish dorsal mesodermal and ectodermal cells are highly elongated and mediolaterally aligned relative to the midline (Concha and Adams, 1998; Topczewski et al., 2001; Jessen et al., 2002; Marlow et al., 2002; Myers et al., 2002b). Typically, imaging of fluorescently labeled cells in live embryos is used to make measurements of cell shape and mediolateral orientation (Topczewski et al., 2001; Myers et al., 2002b). Cell elongation is measured as the length-to-width ratio (LWR), where length and width represent the largest measurable distances along each cellular axis (Fig. 5A). Alternatively, cell elongation can be analyzed in terms of roundness, which is a measure of

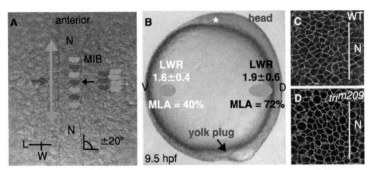

Fig. 5. Strong extension movements with limited convergence characterize the dorsal domain of converg-ence and extension. (A) Mediolateral intercalation behaviors (MIB) underlying anteroposterior extension of axial tissues may be limited to the notochord (N) in zebrafish embryos. (A and B) Cell elongation and mediolateral-orientation are informative cell behaviors to examine in the dorsal domain. Asterisk denotes animal pole. (C and D) In *tri* mutant embryos, paraxial ectodermal cells are rounder and less biased in their orientation than wild-type cells. Cells were labeled with membrane-targeted GFP and live embryos were analyzed using confocal microscopy. (See Color Insert.)

a cell's surface area and volume (Ulrich et al., 2003). Using these tools, it was shown that at late gastrulation (9.5–10 hpf) paraxial ectodermal and mesodermal cells are elongated with LWRs of 1.9 ± 0.6 whereas rounder ventral cells have LWRs of 1.6 ± 0.4 (Fig. 5B; Topczewski et al., 1991; Myers et al., 2002b). A cell is classified as mediolaterally aligned if its long axis is within $\pm 20°$ of a line perpendicular to the embryonic axis as represented by the notochord (Fig. 5A). At late gastrulation, dorsal ectodermal and mesodermal cells are predominantly aligned with the mediolateral axis (>70% of measured dorsal cells are oriented $\pm 20°$; Fig. 5B; Topczewski et al., 2001; Jessen et al., 2002; Myers et al., 2002b). By contrast, rounder ventral cells are less biased with respect to their orientation (40% of measured ventral cells are mediolat-erally aligned; Myers et al., 2002b). It has become clear that proper mediolateral cell elongation is critical for convergence and extension cell movements in the dorsal domain (Fig. 5C and D). Using mosaic analyses to assess the behavior of wild-type cells in a dominant-negative Rho kinase 2 (Rok2) expressing host and vice versa, Marlow et al. (2002) showed that cell elongation can occur independently of medio-lateral alignment. This is an important finding because it supports the notion that, in zebrafish, cell elongation is an autonomous cell behavior without significant influence from neighboring cells, and that additional cues are required for an elongated cell to acquire mediolateral orientation. Moreover, in *trilobite* (*tri*)/*strabismus* and *knypek* (*kny*)/*glypican 4/6* mutant embryos, defective mediolateral cell elongation in dorsal tissues results in compromised axis extension (Topczewski et al., 2001; Jessen et al., 2002). Conversely, ventral mesodermal cells in dorsalized *somitabun*/*smad5* mutants, in which ectopic, dorsally directed, intercalations in ventral mesoderm lead to con-vergence and extension movements around the circumference of the embryo, exhibit LWRs of 1.9 ± 0.5, comparable to dorsal cells (Myers et al., 2002b). The significance of mediolateral cell elongation to convergence and extension movements has been revealed by the observation that disruption of cell morphology in paraxial mesoderm is associated with an inability of cells to move dorsally along straight trajectories (Marlow et al., 2002). This relationship between mediolateral cell elongation and effective movement is also observed in the lateral domain where convergence is driven by directed cell migration (see Section 3.2).

Mediolaterally directed intercalation of bipolar cells might not be the only mode of cell rearrangement occurring in the zebrafish dorsal gastrula domain. In *Xenopus* ectoderm, monopolar protrusive activity drives dorsally directed cell intercalation in a manner dependent upon the underlying mesoderm (Elul and Keller, 2000). Zebrafish dorsal ectodermal cells are mediolaterally elongated and cell groups become dispersed anteroposteriorly, suggesting that extension of this tissue is mediated at least in part through intercalation behaviors. However, it is not clear whether ectodermal cell intercalation involves bipolar and/or monopolar protrusive activity. When considering cell behaviors that shape dorsal tissues, it is important to remember that formation of axial structures such as notochord and somites is concurrent with convergence and extension movements. Therefore, local morphogenetic cell behaviors may vary between adjacent cell populations and germ layers. For example, while bipolar intercalation continues for *Xenopus* notochordal mesoderm, paraxial cells at the notochord-somitic boundary utilize monopolar protrusive activity to grab and pull neighboring cells towards the midline (Shih and Keller, 1992b; Keller et al., 2000). Further characterization of cell behaviors controlling strong extension in the dorsal gastrula domain of zebrafish, and the relationship between these behaviors and notochord/somite formation are important areas for future research. We do not know for example, to what extent mediolateral intercalation behaviors occur in paraxial mesoderm.

3.2. Lateral domain

In the lateral domain of the zebrafish gastrula, dynamic changes in cell behaviors and movement contribute to increasing convergence and extension movements. Tracing movements of cell populations in zebrafish embryos demonstrated that lateral cell groups labeled at the beginning of gastrulation form anteroposteriorly distributed arrays in lateral regions as they move from lateral to dorsolateral regions (Kimmel et al., 1994; Sepich et al., 2000; Myers et al., 2002b). This is in contrast to another teleost *Fundulus*, where lateral gastrula cells strictly undergo convergence without extension until they reach the dorsal region (Trinkaus et al., 1991, 1992). Importantly, current data from zebrafish suggest that dorsal convergence of lateral cells and anteroposterior extension of the embryonic axis can be physically and genetically separable events. For example, whereas defective Nodal signaling in *one-eyed pinhead* mutant embryos inhibits mesendoderm formation and extension of the axis, convergence of cells to the dorsal region appears unaffected (Feldman et al., 1998; Gritsman et al., 1999; Erter et al., 2001). Furthermore, in *tri* mutant embryos, where both convergence and extension movements are reduced, the ability of mutant cell populations to disperse along the anteroposterior axis is at least partially independent of their ability to converge dorsally (Sepich et al., 2000). These data suggest that distinct molecular pathways may operate in different regions of the gastrula to control distinct cell behaviors underlying convergence and extension movements.

In both *Fundulus* and zebrafish, the speed of convergence of lateral cells toward the dorsal axis increases throughout gastrulation (Trinkaus, 1998; Sepich et al., 2000). By measuring the convergence of fluorescently labeled cell populations, Sepich et al. (2000) showed that an initially slower convergence speed increases significantly during mid- to late gastrulation (80% epiboly-yolk plug closure; 8.5–9.5 hpf) and peaks by the end of the gastrula period (yolk plug closure-tailbud; 9.5–10 hpf). Recent studies show that at the beginning of gastrulation, extension occurs without convergence as cells migrate towards the animal pole showing no dorsal bias in their movement trajectories (shield stage-~55% epiboly; 6–7 hpf; Fig. 6A,a; D.S. Sepich and L.S.-K., unpublished data). Convergence begins at ~7.5 hpf and is relatively slow as cells move along complex trajectories (Fig. 6B,b). Extension of the embryonic axis occurs because the trajectories of some cells are biased towards the animal pole while those of others are biased vegetally. From mid-gastrulation to the end of the gastrula period (~7.5–10.3 hpf), cells translocate from the lateral domain of convergence and extension into the dorsal domain where their convergence speed dramatically slows as they approach the dorsal midline (Fig. 6B–D). What are the specific cell behaviors underlying the increased rates of dorsal convergence? Experiments performed in *Fundulus* demonstrated that cells located further from the dorsal axis meander more than cells located dorsally (Trinkaus, 1998). Therefore, these lateral cells exhibit zigzagging trajectories and consequently achieve little net movement towards dorsal. The data have now been corroborated and extended in zebrafish. Using time-lapse Nomarski imaging, Jessen et al. (2002) reported that at mid-gastrulation (80% epiboly; 8.5 hpf) lateral mesodermal cells migrate as individuals along indirect paths with a slow net dorsal speed (Fig. 6b). By the end of the gastrula period however, lateral mesodermal cells migrate dorsally at increased net speeds along more direct trajectories (Fig. 6c and d). This study also demonstrated that the change in trajectories and net speed of movement is associated with a morphological cell transformation. At mid- to late gastrulation lateral mesodermal cells are rounder with an amoeboid-like morphology (Fig. 6b), but by the end of the gastrula period, become significantly more densely packed and mediolaterally elongated (Fig. 6c and 6d; Jessen et al., 2002). In zebrafish then, there is a correlation between the mediolateral elongation of lateral mesodermal cells and the increase in their net dorsal migration speed towards the end of the gastrula period. Collectively, these studies reveal striking similarities between dorsal convergence of lateral cell populations in teleost gastrulae and the chemotactic behaviors of *Dictyostelium* amoebae and leukocytes migrating towards a source of chemoattractant (Chung et al., 2001). Intriguingly, this may indicate that a gradient (high [dorsal] and low [ventral]) of chemoattractant promotes the early dorsally directed migration of lateral cells (Fig. 6). This possibility and candidate signaling molecules will be discussed in Section 4.

Insight into the genetic mechanisms regulating the observed changes in cell morphology and migratory properties in the lateral gastrula domain has been provided through analyses of *tri* mutants. Homozygous *tri* mutant embryos exhibit broader

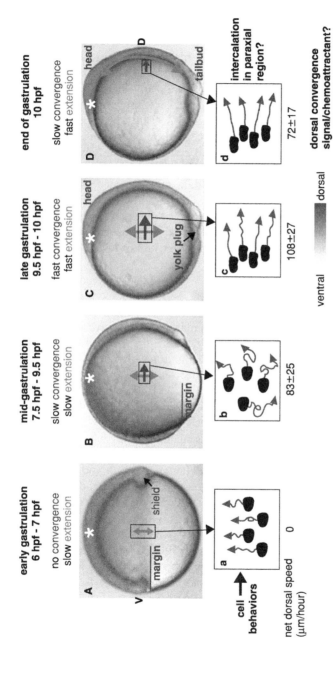

Fig. 6. In the lateral domain, diverse cell behaviors lead to increasing convergence and extension movements as lateral cells approach paraxial and axial regions of the embryo. (A,a) During early gastrulation, lateral cell populations move towards the animal pole (asterisk) as individuals while epiboly movements continue to move the blastoderm vegetally. (B,b and C,c) From mid- to late gastrulation, lateral cells undergo a morphological change, becoming elongated and migrating as densely packed cohorts. (D,d) In paraxial regions, the net dorsal migration speed slows. These cells will contribute to formation of dorsal structures such as somites and the notochord. (See Color Insert.)

and shorter embryonic axes as a result of defective convergence and extension movements (compare Fig. 3B,b and C,c; Hammerschmidt et al., 1996; Solnica-Krezel et al., 1996; Sepich et al., 2000; Jessen et al., 2002). It was shown that the slow convergence of lateral cell populations at early and mid-gastrulation stages is normal in *tri* mutant embryos (Sepich et al., 2000; Jessen et al., 2002). Accordingly, the morphology and net dorsal migration speed of *tri* mutant cells does not differ significantly from wild type at early to mid-gastrulation (Jessen et al., 2002). In contrast to wild type however, *tri* mutant cells do not increase their rate of dorsal convergence at mid- to late gastrulation stages. Notably, *tri* mutant cells remain rounded and continue to migrate dorsally at slower net speeds and along less direct trajectories than their wild-type counterparts. Furthermore, recent data suggest that similar to *tri* mutants, lateral mesodermal cells in *kny* mutant embryos are also rounder than wild type at the end of the gastrula period (D.S. Sepich and L.S.-K., unpublished data). Taken together, these data suggest that *tri*- and *kny*-dependent mediolateral cell elongation is necessary for dorsal migration of lateral mesodermal cells along straight paths and consequently for the increased velocity of their convergence movements. What molecules control the convergence and extension cell movements of lateral tissues during early and mid-gastrulation?

3.3. Ventral domain

In the ventral domain of the blastoderm margin, prospective mesendodermal cells move beneath the epiblast and subsequently engage in anterior migration towards the animal pole. However, in contrast to lateral and dorsal domains of the gastrula, these cells do not participate in convergence and extension movements at mid- and late gastrulation stages (Myers et al., 2002b). Instead, they migrate toward the vegetal pole and contribute to formation of the tailbud at the end of epiboly (Fig. 4C). Cells in this "no convergence/no extension zone" (NCEZ) occupy a 20°–30° arc of the ventral blastoderm margin. This ventral domain overlaps with areas of high bone morphogenetic protein (Bmp) activity. Bmps are members of the TGFβ superfamily and act as ventralizing factors (reviewed by Hammerschmidt and Mullins, 2002). The significance of increased Bmp signaling to specification of the ventral domain of convergence and extension movements became clear through analysis of the highly ventralized *bozozok;dino/chordin* double mutants (Gonzalez et al., 2000). Myers et al. (2002b) demonstrated that in these mutants, where Bmp activity encompasses a larger than normal domain, the NCEZ is expanded to ≥180° (Myers et al., 2002b). Further underscoring the requirement for Bmp signaling in NCEZ specification, NCEZ cell behaviors are absent in dorsalized mutants such as *somitabun/smad5* (Myers et al., 2002b). Time-lapse recordings of the migratory behaviors of NCEZ cells indicate these cells move as individuals along net paths oriented towards the vegetal pole (Myers et al., 2002b). Vegetally migrating mesendodermal cells in the NCEZ meander substantially in the dorsoventral direction, suggesting these cells are being directed towards the vegetal pole and not the dorsal axis. This is in marked contrast to cells in lateral and dorsal convergence and extension domains.

4. Molecular and genetic control of convergence and extension

The number and diversity of cell movement behaviors required for narrowing and elongation of the embryonic axis suggest the involvement of an equally complex network of molecules and signaling pathways. These cellular behaviors, which include directed migration and radial and mediolateral intercalations, must be coordinated in the context of individual cells, cell groups, and epithelial-like cell sheets. Consequently, molecules regulating cell polarity, cytoskeletal assembly and structure, and cell–cell/cell-extracellular matrix adhesion are good candidates for controlling these movements. Furthermore, coincident with gastrulation movements, cells are being specified for different fates depending on their position within the embryo. We know that in zebrafish, the ventral to dorsal Bmp activity gradient, which specifies cell fates within nascent germ layers (reviewed in Hammerschmidt and Mullins, 2002), also controls convergence and extension movements at least partly by inhibiting Wnt11 and Wnt5 gene expression in the ventral gastrula domain (Myers et al., 2002b). While many molecules have been implicated in the control of gastrulation movements including the secreted protein Cyr61 (Latinkic et al., 2003), the receptor tyrosine kinase Ror2 (Oishi et al., 2003), the Slit/Robo (Yeo et al., 2001) and Eph/Ephrin (Oates et al., 1999; Chan et al., 2001) systems, and the integrin/cadherin adhesion molecules (Marsden and DeSimone, 2003), this section is not intended to be a comprehensive review. For additional reviews of the molecules and signaling pathways controlling gastrulation cell movements see Heisenberg and Tada (2002), Keller (2002), Myers et al. (2002a), Solnica-Krezel and Cooper (2002), Veeman et al. (2003a), Wallingford et al. (2002). Here we discuss our current knowledge of three emerging and important areas: (A) noncanonical Wnt signaling in cell polarity and cell movement, including functional interactions between noncanonical and canonical Wnt signaling pathways, (B) Heterotrimeric G proteins and gastrulation cell movements, and (C) Stat3, Phosphoinositide 3-kinase, and chemotactic-like cell behaviors during gastrulation.

4.1. Noncanonical Wnt signaling—cell polarity and beyond

Members of the Wnt protein family are secreted glycoproteins that play diverse roles in a variety of signaling pathways involved in both normal development and disease. Wnt1 was first identified as a proto-oncogene abnormally activated in mammary tumors by integration of mouse mammary tumor virus (Nusse and Varmus, 1982). Subsequent research demonstrated that Wnt signaling involves regulation of a large multiprotein complex containing Axin, APC, and GSK-3. In the absence of a Wnt signal, this complex modifies β-catenin leading to cytosolic destruction by the proteosome (reviewed in Wodarz and Nusse, 1998). β-catenin is the major effector of canonical Wnt signaling, acting downstream of Frizzled and Disheveled as a nuclear transcriptional regulator to control processes such as embryonic induction, cell polarity, cell fate specification, and movement (Fig. 7; reviewed in Cadigan and Nusse, 1997; Cong et al., 2003). The existence of noncanonical, or β-catenin–independent Wnt signaling pathways was first suggested by experiments in *Xenopus*

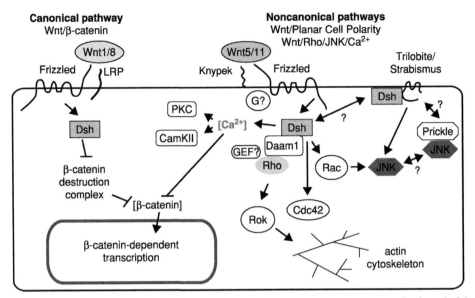

Fig. 7. Schematic diagram of canonical and noncanonical Wnt signaling pathways operating in early fish and frog embryos. Not all molecules and interactions are shown. See text for details.

showing that overexpression of certain Wnts disrupts gastrulation cell movements but not cell fate specification (Moon et al., 1993; Torres et al., 1996). While Wnts represent a single large family of related proteins, their ability to stimulate distinct cellular processes such as transformation and morphogenetic movements has led to the further categorization of Wnts into functional classes. Predominantly, the noncanonical Wnts (Wnt4, Wnt5a, and Wnt11) regulate morphogenesis by influencing diverse intracellular processes such as cytoskeleton remodeling, Ca^{2+} fluxes, and Jun N-terminal kinase (JNK) activity (Fig. 7). It is important to note that while different noncanonical Wnt effectors appear to activate specific molecular pathways, there is evidence suggesting that several of these molecules function in multiple pathways, including canonical Wnt/β-catenin signaling. Therefore, it is not clear whether there are distinct noncanonical Wnt signaling modules or if there is a single pathway whose molecular components have different context-dependent functions (reviewed in Veeman et al., 2003a).

In zebrafish and *Xenopus*, a noncanonical Wnt pathway regulates cell morphology underlying both migration and intercalation behaviors of individual cells, cell groups, and cell sheets during gastrulation. This signaling pathway uses components similar to the *Drosophila melanogaster* planar cell polarity (PCP) pathway that controls the polarization of different cell types within the plane of several epithelial tissues including the wing and eye. For example, PCP signaling in the wing directs the proximal-distal polarization of epithelia such that each cell produces a single, distally pointed hair (reviewed in Adler, 2002). For detailed comparisons of *Drosophila* and vertebrate PCP pathways, see Mlodzik (2002) and Strutt (2003). Here we discuss components of the vertebrate PCP pathway identified in zebrafish and *Xenopus* and how they affect cell behaviors underlying convergence and extension movements. We also discuss recent genetic data suggesting that noncanonical Wnts can antagonize canonical Wnt/β-catenin signaling.

Much of our knowledge of PCP signaling in vertebrates comes from analyses of zebrafish mutants with defective gastrulation cell movements. In zebrafish, *silberblick* (*slb*) mutations disrupt the Wnt11 gene and cause defects in convergence and extension of dorsal mesendodermal and neuroectodermal tissues (Heisenberg et al., 1996; Heisenberg and Nüsslein-Volhard, 1997; Heisenberg et al., 2000). A recent study by Ulrich et al. (2003) has provided important insight regarding the cellular mechanisms by which Wnt11 regulates movements during gastrulation. These authors showed that in *slb*/Wnt11 mutant embryos, mesendodermal cells exhibit defects in both the directionality and velocity of anteriorward migration at the onset of gastrulation. Defective anterior migration of mesendodermal cells in *slb* mutants was correlated with the inability of these cells to orient their cellular processes along the direction of migration. Interestingly, the *slb* mutant cells had cellular processes that were comparable to wild type in number, shape, and length (Ulrich et al., 2003). Consistent with these sites of action, *slb*/Wnt11 is expressed in anterior paraxial mesendoderm and lateral neuroectoderm at the end of the gastrula period (Makita et al., 1998; Heisenberg et al., 2000). By contrast, Wnt5 transcripts are predominantly detected within more posterior paraxial and axial mesendoderm (Kilian et al., 2003). The Wnt5 protein, encoded by the *pipetail* (*ppt*) locus (Hammerschmidt et al., 1996; Rauch et al., 1997), has been implicated in the regulation of convergence and extension cell movements in both anterior and posterior regions of the gastrula (Kilian et al., 2003). Moreover, analysis of *slb;ppt* double mutant embryos demonstrated that Wnt11 and Wnt5 function redundantly in anterior mesendoderm (Kilian et al., 2003). It was shown that *ppt*/Wnt5 regulates mediolateral cell elongation and orientation of cellular processes in posterior ectoderm and mesendoderm (Kilian et al., 2003). In the picture emerging from these studies, Wnt11 regulates the movement of anterior mesendodermal and lateral neuroectodermal cells while Wnt11 and Wnt5 have partially redundant roles in convergence and extension of more posterior mesendodermal and ectodermal tissues. However, we still do not know whether Wnt11 and Wnt5 actively direct cell movements or rather, act permissively to create an environment favorable to cell movement. In support of the latter, neither Wnt5 nor Wnt11 mRNA is expressed (protein distribution remains unknown) in a graded manner in zebrafish gastrulae, and ubiquitous overexpression of Wnt11 can suppress both the *slb* cell morphology and movement phenotypes (Ulrich et al., 2003).

Frizzled (fz) proteins are members of a large family of seven transmembrane-domain cell surface receptors for both canonical and noncanonical Wnt ligands (Bhanot et al., 1996). Biochemical, epistatic, and expression analyses in zebrafish and *Xenopus* have suggested that Fz7 and Fz2 function as downstream mediators of Wnt11 and Wnt5, respectively (Djiane et al., 2000; Sumanas et al., 2001; Kilian et al., 2003). Furthermore, data from *Xenopus* indicate that Fz7 functions in a variety of developmental contexts to control embryonic patterning and morphogenesis through both Wnt/β-catenin and noncanonical Wnt pathways, but whether all of these effects is mediated by Wnt11 is unclear (Djiane et al., 2000; Medina et al., 2000; Sumanas et al., 2000; Sumanas and Ekker, 2001; Winklbauer et al., 2001). Reception of Wnt11 and Wnt5 ligands at the plasma membrane is thought to be facilitated by the glypican-class of heparan sulfate proteoglycans. In zebrafish, *kny/glypican 4/6* mutant embryos exhibit defects in convergence and extension cell movements due at least in part to impaired mediolateral cell elongation (Topczewski et al., 2001). This study showed that *kny* genetically interacts with *slb*/Wnt11 and is capable of potentiating the ability of Wnt11 to suppress the *slb* mutant phenotype. In contrast to Wnt11, *kny* is widely expressed during embryogenesis with transcripts enriched dorsally in the early gastrula (Topczewski et al., 2001). Similarly, in *Xenopus* Glypican 4 regulates gastrulation movements by influencing noncanonical Wnt signaling (Ohkawara et al., 2003). Molecularly, Glypican 4 binds Wnt11, Wnt5A, Wnt8, and Frizzled7 *in vitro* and influences the membrane association of the cytoplasmic protein Disheveled in dorsal mesoderm (Ohkawara et al., 2003).

Disheveled (Dsh) is a modular cytoplasmic protein that acts downstream of *fz* receptors in both Wnt/β-catenin and noncanonical Wnt signaling pathways (reviewed in Boutros and Mlodzik, 1999). Dsh possesses three highly conserved domains termed DIX (*Di*sheveled, *Ax*in), PDZ (*P*ost synaptic density-95, *D*iscs-large, and *Z*onula occludens-1), and DEP (*D*isheveled, *E*gl-10, and *P*leckstrin) as well as several other conserved regions (reviewed in Boutros and Mlodzik, 1999 Wharton, 2003; Penton et al., 2002). Largely from work in *Xenopus*, we learned that the Dsh DIX domain is important for Wnt/β-catenin signaling while the DEP and PDZ domains are required for gastrulation movements (Tada and Smith, 2000; Wallingford et al., 2000). Wallingford et al. (2000) provided the first evidence that Dsh mediates cell elongation underlying convergent extension movements in *Xenopus* embryos. This study further showed that loss of Dsh function affects the ability of dorsal mesodermal cells to stabilize and mediolaterally orient their protrusions (Wallingford et al., 2000). Moreover, while ectopic expression of Dsh inhibits convergent extension, dorsal mesodermal cells overexpressing Dsh do not exhibit clear defects in their membrane protrusions (Wallingford et al., 2000). From work in *Drosophila*, it is clear that recruitment of Dsh to the plasma membrane is necessary for the PCP pathway but not canonical Wnt (Wingless) signaling (Axelrod et al., 1998). In *Xenopus*, Fz7 signaling promotes Dsh phosphorylation and translocation to the membrane (Rothbächer et al., 2000), and Dsh is localized to the membrane in cells engaged in convergent extension movements (Wallingford et al., 2000). In zebrafish, the *slb*/Wnt11 mutant phenotype is suppressed by overexpression of a truncated form of Dsh (lacking the DIX domain) that is thought to transduce morphogenetic, but not Wnt/β-catenin signaling (Heisenberg et al., 2000). Taken together, the current data suggest that noncanonical Wnts effect changes in cell movement behaviors by modulating the cytoskeleton via Dsh function.

How does Dsh communicate with the actin cytoskeleton? Habas et al. (2001) showed that Wnt/Fz signaling stimulates formation of a complex that includes Dsh, the small GTPase RhoA, and the novel Formin homology protein Daam1. This complex activates RhoA through an unidentified, but presumably associated Rho guanine nucleotide exchange factor. Numerous target proteins downstream of activated Rho have been identified in cell culture systems (reviewed in Schmitz et al., 2000). Many of these proteins are linked to actin reorganization such as Rho-associated kinase (Rok) and members of the Diaphanous formin subfamily (reviewed in Ridley, 2001). Experiments in zebrafish have demonstrated that Rok2 acts downstream of Slb/Wnt11 to control mediolateral cell elongation and dorsal movement of mesodermal cells along straight paths (Marlow et al., 2002). In addition to Rho/Rok activation, Habas et al. (2003) showed that Wnt/Fz signaling in *Xenopus* also activates the small GTPase Rac through a mechanism that is both dependent on Dsh/Daam1 and independent of the Wnt-induced Dsh/RhoA complex. The Dsh/Rac complex is necessary both for the activation of JNK and proper gastrulation movements (Habas et al., 2003). The relationship between JNK activity and cell movement behaviors is unknown, though it has been suggested that noncanonical Wnt signaling (Torres et al., 1996) and JNK (Yamanaka et al., 2002) are capable of influencing cell–cell adhesion. While the ability of Wnt/Fz signaling to activate Rho and Rac demonstrates how Wnts may communicate with the cytoskeleton it must be noted that Dsh and Rok2 are the only cytoplasmic proteins identified thus far that mediate Wnt/Fz signaling in zebrafish embryos. By investigating how RhoA and Rac1 impact cell behaviors in *Xenopus* gastrulae, Tahinci and Symes (2003) have provided insight into how these small GTPases control morphogenetic movements. Interestingly, these authors found that Rho and Rac behave somewhat differently *in vivo* than in previously reported *in vitro* assays (Tahinci and Symes, 2003). Both Rho and Rac influence the direction and activity of membrane protrusions and formation of lamellipodia. However, Rho is distinctly required for the mediolateral elongation of dorsal mesodermal cells undergoing intercalation behaviors, while Rac is necessary for filopodia formation. Taken together, the works of Habas et al. (2001, 2003) and Tahinci and Symes (2003) demonstrate that Wnt/Fz signaling diverges at Dsh leading to the activation of both Rho and Rac, and diverse changes in cell movement behaviors presumably through modulation of cytoskeletal architecture.

Two proteins shown to be critical for PCP in *Drosophila* and gastrulation cell movements in zebrafish and *Xenopus* are the putative transmembrane protein Strabismus/Van Gogh (Stbm/Vang) and the cytoplasmic PET/LIM domain protein Prickle (Pk). Mutations in the zebrafish *tri* locus disrupt a *stbm/vang* homologue that likely is the orthologue of mouse Loop-tail associated protein/Lpp1 and human Strabismus1/Vangl2 (Jessen et al., 2002; Park and Moon, 2002; J.R.J. and L.S.-K., unpublished data). Homozygous *tri* mutant embryos exhibit defects in mediolateral cell elongation underlying effective convergence and extension movements and directed cell migration (Sepich et al., 2000; Jessen et al., 2002). In the loss-of-function situation (*tri/stbm* mutant cells), *tri/stbm* functions both cell autonomously and nonautonomously to control elongation and mediolateral alignment of dorsal gastrula cells (Jessen et al., 2002). In contrast, analyses of ectopic *Xenopus stbm* function in dorsal explants suggested only a nonautonomous role for *stbm* in controlling mediolateral cell elongation and protrusive activity (Goto and Keller, 2002). The ability of *Xenopus stbm* overexpressing cells to be rescued by the wild-type environment indicates that, unlike loss of *stbm* function, these cells are competent to receive extracellular information and execute a cell polarity program. Wild-type embryos injected with *pk* morpholino antisense nucleotides exhibit convergence and extension defects that are milder but similar to *tri* mutant embryos (Carreira-Barbosa et al., 2003; Veeman et al., 2003b). The *pk* morpholino-injected embryos exhibit a posteriorly shifted prechordal plate, a broader neural plate, and a shorter and wider notochord, indicative of cell movement defects (Fig. 3D). Overexpression of *pk* in wild-type embryos causes a similar defect in convergence and extension (Carreira-Barbosa et al., 2003). Interestingly, while *tri/stbm* is widely expressed before and throughout gastrulation in each germ layer (Park and Moon, 2002; J.R.J. and L.S.-K., unpublished data), zygotic *pk* expression initiates on the dorsal side of the embryo before becoming broadly expressed in mesendodermal and neuroectodermal tissues (Carreira-Barbosa et al., 2003; Veeman et al., 2003b).

How do Tri/Stbm and Pk function to mediate gastrulation cell movements? Both zebrafish Tri/Stbm and *Xenopus* Pk were shown to bind Dsh (Park and Moon, 2002; Takeuchi et al., 2003) and in *Drosophila*, Stbm/Vang and Pk were found to interact physically (Bastock et al., 2003). In addition, Tri/Stbm and Pk are capable of influencing JNK activity *in vitro* and Pk binds JNK *in vitro* (Park and Moon, 2002; Takeuchi et al., 2003; Veeman et al., 2003b). From work in *Drosophila* we know that *stbm/vang* and *pk* interact genetically with each other and with other components of the PCP pathway (reviewed in Strutt, 2003). However, these interactions are complex and efforts to place *stbm/vang* and *pk* within the proper context of *Drosophila* PCP signaling have been unsuccessful. In zebrafish, *tri/stbm* genetically interacts with *kny/glypican 4/6* (Marlow et al., 1998) and *slb*/Wnt11 (Heisenberg and Nüsslein-Volhard, 1997), and similarly, injection of *pk* morpholino into *tri* homozygous mutant and heterozygous embryos enhances the gastrulation phenotype (Carreira-Barbosa et al., 2003). The current data indicate that Tri/Stbm and Pk do not function as simple positive or negative regulators of Wnt/PCP signaling during gastrulation (Jessen et al., 2002; Carreira-Barbosa et al., 2003). A key area of future research will be to determine the functional consequence of interactions between Tri/Stbm and Pk, and Dsh, and whether these proteins form a ternary complex and co-localize at the plasma membrane. Overexpression of *stbm* in *Xenopus* animal caps and in zebrafish blastula cells results in translocation of Dsh to the plasma membrane (Park and Moon, 2002; J.R.J. and L.S.-K., unpublished data). There is conflicting data regarding the ability of Pk to influence Dsh membrane localization in *Drosophila* (Tree et al., 2002; Bastock et al., 2003), and zebrafish and *Xenopus* (Carreira-Barbosa et al., 2003; Takeuchi et al., 2003; Veeman et al., 2003b). In perhaps the simplest scenario, Tri/Stbm and Pk may cooperate to antagonize Dsh thereby limiting Wnt/Fz signaling to specific subcellular domains. In support of this notion, Kinoshita et al. (2003) have employed tagged proteins to demonstrate that *Xenopus* Dsh, PKCδ, Rac, and Arp3 are localized at the tips of elongated mesodermal cells in dorsal explants.

Given the significant role of Dsh in Wnt/β-catenin signaling, the ability of Tri/Stbm and Pk to bind Dsh suggests that canonical and noncanonical Wnt pathways may be antagonistic. Although this concept has gained scattered support for many years (Torres et al., 1996), there is little data regarding the physiological relevance of this antagonism. Notably, disruption of Wnt/PCP signaling by gain- or loss-of-function does not appear to affect canonical Wnt/β-catenin signaling *in vivo*, as typically measured by marker gene expression (Habas et al., 2001, 2003; Topczewski et al., 2001; Darken et al., 2002; Goto and Keller, 2002; Jessen et al., 2002; Marlow et al., 2002; Carreira-Barbosa et al., 2003; Ohkawara et al., 2003; Takeuchi et al., 2003; Veeman et al., 2003b). In zebrafish and *Xenopus*, the canonical Wnt/β-catenin pathway plays important roles in dorsal axis formation and anteroposterior neural patterning and operates concurrently with noncanonical Wnt signaling in cells undergoing gastrulation cell movements. Therefore, it can be argued that either noncanonical Wnt signaling does not influence Wnt/β-catenin signaling or the influence during gastrulation is subtle, such that it does not produce phenotypic effects and requires sensitive methods of detection. In the preceding paragraphs, we discussed evidence implicating Wnt11 and Wnt5 in the regulation of convergence and extension cell movements via a vertebrate Wnt/PCP pathway. However, these and other noncanonical Wnt ligands can also signal through a Wnt/Ca^{2+} pathway (reviewed in Kühl et al., 2000b) involving Ca^{2+} fluxes (Slusarski et al., 1997a,b), protein kinase C (PKC; Sheldahl et al., 1999), Ca^{2+}/calmodulin-dependent protein kinase II (CamKII; Kühl et al., 2000a), and the small GTPase Cdc42 (Choi and Han, 2002). Besides its role in PCP signaling, Dsh was also proposed to act downstream of the putative Wnt11 receptor Fz7 to stimulate Ca^{2+} release and PKC and CamKII activity (Fig. 7; Sheldahl et al., 2003). Recent data have now provided genetic evidence that Ppt/Wnt5 functions in the Wnt/Ca^{2+} pathway to antagonize Wnt/β-catenin signaling and promote formation of ventral cell fates (Westfall et al., 2003). These authors demonstrated that overexpression of zebrafish noncanonical Wnt4, Ppt/Wnt5, and Slb/Wnt11 induces the release of intracellular Ca^{2+}. Moreover, gain of Wnt5 function causes hyperventralization of zebrafish embryos as revealed morphologically at day one of development and by loss of dorsal gene expression at the gastrula stage (Westfall et al., 2003). Conversely, these authors showed that *ppt* mutant embryos lacking both maternal and zygotic wild-type Wnt5 function exhibit dorsalized phenotypes with an expansion of dorsal tissues. Increased amounts of nuclear β-catenin and up-regulation of the β-catenin target gene *bozozok* (Fekany et al., 1999; Ryu et al., 2001; Leung et al., 2003) are consistent with the notion that canonical Wnt signaling causes the maternal-zygotic *ppt* mutant phenotype (Westfall et al., 2003). It is important to note that previous studies utilizing different (probably weaker?) *ppt* alleles, and wild-type embryos injected with a morpholino directed against *ppt* did not identify dorsal-ventral patterning defects (Hammerschmidt et al., 1996; Rauch et al., 1997; Lele et al., 2001; Kilian et al., 2003). Nevertheless, the study by Westfall et al. (2003) raises several important questions regarding molecular interactions not only between noncanonical and canonical Wnt pathways, but also between the noncanonical Wnt/PCP and Wnt/Ca^{2+} pathways themselves. Do these two noncanonical Wnt pathways overlap as some evidence suggests (Veeman et al., 2003b; Westfall et al., 2003), or do they operate separately, controlling distinct aspects of cell movement? How are the Wnt/PCP and Wnt/Ca^{2+} pathways regulated in dorsoposterior mesendodermal and ectodermal cells, where Wnt11 and Wnt5 simultaneously regulate convergence and extension cell movements (Lele et al., 2001; Kilian et al., 2003). Important future experiments will identify the specific Fz receptors and downstream factors stimulated by Wnt11 and Wnt5 *in vivo*, and the cell movement behaviors they control. Further analysis of Dsh and its binding partners (e.g., Tri/ Stbm, Pk, PKC, Daam1) will also provide new insight into how this modular protein toggles Wnt signaling through both canonical and noncanonical pathways.

4.2. Heterotrimeric G proteins—downstream or parallel to noncanonical Wnt signaling?

Fz proteins resemble members of a very large family of G-protein-coupled receptors characterized by their seven membrane-spanning regions. Ligand binding to these receptors stimulates conformational changes that expose heterotrimeric G protein binding sites in the intracellular domain. Subsequently, GDP is exchanged for GTP on the G protein α subunit. The released $G\alpha$ or $G\beta\gamma$ subunits interact with a variety of effector molecules to initiate diverse intracellular signaling responses. $G\alpha$ subunits are divided into four families based on their sequence similarity: α_s, α_i, α_q, and α_{12} (Wilkie et al., 1992). Multiple β and γ subunits have also been identified. Slusarski et al. (1997a) provided provocative evidence suggesting that heterotrimeric G proteins mediate noncanonical Wnt/Ca^{2+} signaling via fz receptors. Since this report, it was shown that activation of Fz7, PKC, and CamKII downstream of noncanonical Wnt signaling may also be G-protein-dependent (Sheldahl et al., 1999; Kühl et al., 2000; Winklbauer et al., 2001). Does this data suggest that noncanonical Wnt/Ca^{2+} and Wnt/PCP signaling are both dependent on G-protein-coupled Fz receptors (Fig. 7)? Alternatively, is signaling via G-protein-coupled receptors restricted to distinct pathways to control cell movement behaviors in specific embryonic tissues? Recent data by Penzo-Mendèz et al. (2003) demonstrate that Wnt11 signaling in *Xenopus* activates Cdc42 in a manner that can be inhibited by coexpression of Pertussis toxin, $G\alpha_{i2}$, or $G\alpha_t$, suggesting a role for $G\beta\gamma$ signaling in this process. Mechanistically, inhibition of $G\beta\gamma$ signaling by these molecules suppresses the disruption of morphogenetic cell movements in *Xenopus* animal cap explants caused by overexpression of Wnt11, $fz7$, or Dsh (Penzo-Mendèz et al., 2003). It was also shown that overexpression of bovine $G\beta1$ and $G\gamma2$ subunits activates Cdc42 through a PKC-dependent mechanism (Penzo-Mendèz et al., 2003). Furthermore, inhibition of $G\beta\gamma$ signaling in the *Xenopus* dorsal marginal zone causes defects in convergent extension cell movements and impairs involution of presumptive mesendoderm (Penzo-Mendèz et al., 2003). However, given that normal convergence and extension movements are necessary for involution (Shih and Keller, 1992b), this may be a secondary defect. While we know that Dsh functions downstream of both the Wnt/Ca^{2+} and Wnt/PCP pathways, is there evidence linking Dsh with G-protein-coupled Fz signaling? Sheldahl et al. (2003) showed that stimulation of Ca^{2+} release by a PCP-specific form of Dsh is insensitive to Pertussis toxin suggesting either an upstream or no requirement for G_i family-dependent Dsh activation. The observation that inhibition of $G\beta\gamma$ signaling suppresses the convergent extension defect caused by overexpression of *Xenopus* Dsh (Penzo-Mendèz et al., 2003) suggests there is cross talk between these molecules. However, it is unclear whether this represents a direct interaction or is the result of associations between Dsh and PKC (Kinoshita et al., 2003; Sheldahl et al., 2003). For example, PKCδ activity is required both for Dsh phosphorylation and its recruitment to the plasma membrane (Kinoshita et al., 2003).

Taken together, the current data suggest that G-protein-dependent signaling may function in noncanonical Wnt/Ca^{2+} signaling to regulate morphogenesis. It is now important to identify biochemical interactions between specific G proteins and Fz receptors and between G protein subunits and downstream signaling molecules, including components of the Wnt/PCP pathway.

Given the evidence implicating Rho family GTPases downstream of Wnt/Fz signaling (Habas et al., 2001, 2003; Choi and Han, 2002; Penzo-Mendèz et al., 2003), guanine nucleotide exchange factors (GEFs) are good candidate molecules for linking heterotrimeric G proteins with small GTPases. Notably, G$\alpha_{12/13}$ binds certain Rho-specific GEFs via their regulator of G protein signaling-like domains resulting in both membrane recruitment and GEF activation (Hart et al., 1998; Bhattacharyya and Wedegaertner, 2000; Fukuhara et al., 2000; Wells et al., 2002). Indeed, Habas et al. (2001) have proposed that a Rho GEF may act within the Dsh/Daam1/RhoA protein complex to facilitate RhoA activation downstream of Wnt11/Fz signaling. Besides uncovering biochemical interactions between G proteins and the Wnt pathway, experiments must also address the specific gastrulation cell movement behaviors affected by G-protein-coupled receptor signaling. From the work on *Xenopus* discussed above, we know that certain G protein subunits can influence overall convergence and extension of the embryo; however, we do not know whether G proteins act specifically or play a general role in cell motility. This issue has been addressed by recent studies showing that in zebrafish, G$\alpha_{12/13}$ signaling influences directed cell migrations during both epiboly and convergence and extension (F. Lin, H. Hamm, and L.S.-K., unpublished data). Intriguingly, these data suggest that during convergence and extension G$\alpha_{12/13}$ function is required for mediolateral cell elongation already at midgastrulation, earlier than defects observed in noncanonical Wnt signaling mutants, *tri/stbm* and *kny/glypican 4/6* (Sepich et al., 2000; Topczewski et al., 2001; Jessen et al., 2002). Furthermore, neither stimulation nor inhibition of Gα_{13} signaling suppresses the *slb*/Wnt11 gastrulation defects (F. Lin, H. Hamm, and L.S.-K., unpublished data), perhaps indicating that Gα_{13} functions parallel to the Wnt/PCP pathway to control at least some cell movement behaviors. What role might G$\alpha_{12/13}$ signaling play in directed cell migration? In other systems, migrating cells exhibit a polarized morphology with defined "frontness" and "backness" characterized by the asymmetric distribution of signaling molecules and cytoskeletal assemblies (reviewed in Firtel and Meili, 2000; Devreotes and Janetopoulos, 2003). It has been proposed that neutrophil-like cells control cell polarity by utilizing distinct G-protein-coupled receptor signaling pathways. In this model, Gα_i mediates "frontness" by generating pseudopodial membrane protrusions at the leading edge via a pathway involving phosphoinositol lipids, Rac, and formation of filamentous actin (Xu et al., 2003). In the "backness" pathway, Gα_{12} and Gα_{13} signaling stimulates downstream activation of Rho, Rho kinase, and myosin leading to contraction at the back of the cell (Xu et al., 2003). Importantly, each of these pathways inhibits the other thereby demarcating distinct cell membrane domains with a protrusive leading edge, facing the chemoattractant, and smoother back and sides that are enriched with myosin and are relatively insensitive to the attractant (Xu et al., 2003). It will be important to ask whether the zebrafish Gα_{12} and Gα_{13} proteins play similar roles to control cell migration during gastrulation? Moreover, can the noncanonical Wnt signaling pathway be incorporated into this model of cell motility?

4.3. Stat3 and Phosphoinositide 3-kinase—chemotactic pathways during gastrulation?

In the above sections we discussed how different components of noncanonical Wnt/PCP and Wnt/Ca^{2+} pathways influence morphogenetic cell movements during gastrulation. The data indicate that many of these molecules regulate mediolateral cell elongation and the organization/localization of cell membrane protrusions. While it is clear that proper cell morphology and protrusive activity are necessary for cell movement, it must be noted that convergence and extension and anterior migration do occur albeit slower in zebrafish mutants such as *tri*, *kny*, and *slb*. Therefore, other possibly redundant signaling molecules and pathways must also shape the gastrula, and indeed, as mentioned in the beginning of Section 4, several have been identified. It has been hypothesized that secreted factors, other than noncanonical Wnts, may play an important role during gastrulation, acting to coordinate both the directed migration of lateral cells towards the dorsal axis and the anterior movement of dorsal mesendodermal cells after internalization (Heisenberg and Tada, 2002; Myers et al., 2002a; Solnica-Krezel and Cooper, 2002; Yamashita et al., 2002). During each of these processes cells exhibit chemotactic-like behaviors (reviewed in Weiner, 2002), migrating along straighter trajectories as they approach their target. Stat3 (signal transducer and activator of transcription 3) has emerged as a candidate regulator of the hypothesized secreted molecule(s) that mediates chemotactic-like cell movements during gastrulation (reviewed in Myers et al., 2002a; Yamashita et al., 2002). Embryos injected with a Stat3 morpholino exhibit a mispositioned head and a shorter and broader embryonic axis consistent with a severe reduction of convergence and extension movements (Yamashita et al., 2002). It was shown that Stat3 functions cell autonomously to control anterior migration of axial mesendodermal cells, but its activity is required nonautonomously in dorsal cells of the shield for dorsal migration of lateral cells (Yamashita et al., 2002). This latter result implies that a secreted factor may be activated downstream of Stat3 in dorsal tissues, perhaps directing the migration of lateral and paraxial cells. Interestingly, Stat3 is activated by phosphorylation downstream of Wnt/β-catenin signaling in a manner dependent upon the Tcf transcription factor, but independent of known β-catenin target genes *bozozok* and the nodal-related *squint* that are involved in axis specification (Yamashita et al., 2002). These tantalizing results pose many questions. Perhaps most important is the identification and analysis of molecules acting downstream of Stat3 and determination of their ability to influence cell migration. Candidate molecules that may mediate cell–autonomous functions of Stat3 include the Rac1 small GTPase (Simon et al., 2000) and the SDF-1/CXCR4 chemokine system (Vila-Coro et al., 1999).

It is known that lipid products of phosphoinositide 3-kinase (PI3K) play important roles during chemotaxis of neutrophils (Servant et al., 2000), *Dictyostelium*

(Jin et al., 2000; Meili et al., 1999), and fibroblasts (Haugh et al., 2000). Montero et al. (2003) have now implicated PI3K in the anterior migration of mesendo-dermal cells, providing compelling evidence that chemotactic-like cell behaviors operate in zebrafish gastrulae. Inhibition of PI3K function causes defects in both the morphology and movement of internalized mesendodermal cells (Montero et al., 2003). Notably, these cells are less elongated and exhibit significantly fewer membrane processes than control embryos. Similar to chemotaxis in other systems (Funamoto et al., 2001; Watton and Downward, 1999), protein kinase B (PKB, also known as Akt) was shown to act downstream of PI3K signaling in zebrafish embryos, becoming co-localized with actin at the leading edge of anterior migrating mesendodermal cells (Montero et al., 2003). Enrichment of PKB at the leading edge may stimulate the localized reorganization of actin and/or regulate cell adhe-sion at specific cell attachment points (Watton and Downward, 1999; Jimenez et al., 2000). Earlier studies in *Xenopus* have suggested that Platelet Derived Growth Factor (PDGF) signaling via PI3K influences the adhesive properties of mesodermal cells during gastrulation (Ataliotis et al., 1995; Symes and Mercola, 1996).

Montero et al. (2003) also demonstrated that PDGF acts upstream of PI3K to regulate cell polarization and formation of cellular processes. In tissue culture, zebrafish mesendodermal cells direct their migration and the orientation of their processes towards a source of PDGF, suggesting that PDGF is sufficient to direct cell migration (Montero et al., 2003). However, given that *pdgf* mRNA is ubiquitous-ly expressed (protein distribution remains unknown) throughout the zebrafish gas-trula (Liu et al., 2002), PDGF itself may play a more permissive role during anterior migration of mesendodermal cells. This notion is corroborated by the fact that although inhibition of PI3K function reduces cell polarization and outgrowth of cellular processes, the ability of mesendodermal cells to direct their migration ante-riorly is not affected (Montero et al., 2003). This is in agreement with work on *Dictyostelium* where inhibition of PI3K disrupts chemotaxis, but does not completely eliminate directed migration towards a chemoattractant (Funamoto et al., 2001). As discussed by Montero et al. (2003), the ability of PI3K-deficient mesendodermal cells to migrate anteriorly calls into question the function of cell polarization and dynamic changes in membrane protrusive activity. It is clear that changes in these parameters are necessary for efficient anterior migration of mesendodermal cells along straight paths, but what is the signal directing this migration? It may be that the PDGF/PI3K pathway interacts with either Stat3 (Yamashita et al., 2002) or noncanonical Wnt signaling (Ulrich et al., 2003) to regulate slightly different aspects of anterior migra-tion and therefore compensatory effects are observed when one pathway is disrupted. This notion is supported by the observation that whereas Slb/Wnt11 is required for proper orientation of cellular processes in the anterior mesendoderm (Ulrich et al., 2003), PDGF/PI3K signaling is needed for the formation of these processes (Montero et al., 2003). Future experiments are needed to identify additional chemotactic pathways in zebrafish gastrulae and the cell movement behaviors they control, and to determine the relationship between these pathways and noncanonical Wnt signaling.

5. Summary and future directions

In this chapter, we have discussed our current knowledge of the morphogenetic cell movements and underlying molecular mechanisms implicated in shaping the zebrafish body plan during gastrulation. We have drawn on data collected from both zebrafish and *Xenopus*, highlighting similarities and differences between these organisms. The picture emerging is one where diverse morphogenetic processes are driven by multiple cell movement behaviors that in turn are controlled by several signaling pathways interacting at many levels. This complexity has created an abundance of data that as yet has only been used to synthesize simplistic models. Further experiments will be needed to address the many outstanding questions that remain. For example, work on *Drosophila* has demonstrated that several components of the PCP pathway become asymmetrically distributed relative to the proximal-distal axis of the wing and that this accounts for the formation of a single hair on the distal cell surface (reviewed in Strutt, 2003). By contrast, few components of the vertebrate noncanonical Wnt/PCP or Wnt/Ca^{2+} pathways are known to have asymmetric distribution in cells engaged in convergence and extension movements. Given the dynamic nature of these cell movements, it may be difficult to detect differences in protein localization. However, asymmetric protein localization does seem likely, as it may explain the observation that gain- and loss-of-function of most PCP molecules causes defective cell movements. Localization of *tri/stbm* and *pk* to anterior and posterior cell surfaces could restrict Wnt/*fz* signaling to the medial and lateral edges of elongated cells engaged in intercalation behaviors. This would help explain the complicated epistatic data suggesting that Tri/Stbm and Pk promote cell movements behaviors by antagonizing Wnt/PCP signaling.

The identification and analysis of signaling pathways regulating gastrulation cell movements will continue to be an important area of research. More biochemical data is needed to connect extracellular signals such as Wnts with intracellular responses such as changes in the cytoskeleton. For example, in zebrafish we do not know whether Wnt/Fz signaling stimulates formation of a Dsh/Daam/Rho complex upstream of Rok, and almost nothing is known about downstream events. Moreover, we have just begun to understand how different signaling pathways communicate with each other to elicit specific cell behaviors. While these molecular experiments are valuable, it remains absolutely critical that we continue to dissect the morphogenetic processes underlying convergence and extension of the embryonic axis. Only with the knowledge of specific cell movement behaviors can we assign the correct cellular readout to noncanonical Wnt, Stat3, PDGF/PI3K, and other signaling pathways. Indeed, there is an abundance of reports demonstrating how different molecules influence global changes in embryonic shape, but few that link these molecules to specific cell behaviors including directed migration and intercalation, or specific cell properties such as adhesion, morphology, and protrusive activity (Table 1). Since embryos having the appearance of a convergence and extension defect can be generated through nonspecific means, it is important that specific roles of any given molecule in this process are assessed. Nevertheless, great strides are being made towards understanding the cellular and molecular control of the morphogenetic

Table 1
Molecules controlling cell migration, polarity, and intercalation in fish and frog embryos

Gene	Organism	Structural homologies	Morphogenetic movements	Cellular functions	References
Wnt11 (*silberblick*)	Zebrafish	Secreted cysteine-rich glycoprotein	Anterior migration, Convergence and extension	Directionality and velocity of cell migration, orientation of membrane protrusions	Heisenberg et al., 2002; Ulrich et al., 2003
Wnt5 (*pipetail*)	Zebrafish	Secreted cysteine-rich glycoprotein	Convergence and extension	Mediolaterally-oriented cell elongation, orientation of membrane protrusions	Kilian et al., 2003
glypican 4/6 (*knypek*)	Zebrafish *Xenopus laevis*	Glypican family of heparan sulfate proteoglycans GPI-anchored	Convergence and extension	Mediolaterally-oriented cell elongation, velocity of cell migration	Topczewski et al., 2001 Ohkawara et al., 2003
g12/13	Zebrafish	Heterotrimeric G proteins	Epiboly, Convergence and extension	Mediolaterally-oriented cell elongation, velocity of cell migration	F.Lin, H. Hamm, L.S.-K., unpublished data
strabismus (*trilobite*)	Zebrafish *Xenopus laevis*	Four pass transmembrane receptor C-terminal PDZ-domain binding motif	Convergence and extension	Mediolaterally-oriented cell elongation, intercalation, directionality and velocity of cell migration	Goto and Keller, 2002; Jessen et al., 2002

(*continued*)

Table 1 (*continued*)

Gene	Organism	Structural homologies	Morphogenetic movements	Cellular functions	References
disheveled	*Xenopus laevis*	DIX, DEP, PDZ, and proline-rich domains	Convergent extension	Mediolaterally-oriented cell elongation, stability and orientation of membrane protrusions	Wallingford et al., 2000
rho kinase 2	Zebrafish	Coiled coil α-helix, kinase, Rho-binding, and pleckstrin homology domains	Convergence and extension	Mediolaterally-oriented cell elongation, directionality and velocity of cell movement	Marlow et al., 2002
rhoA	*Xenopus laevis*	Rho family small GTPase	Convergent extension	Mediolaterally-oriented cell elongation, intercalation, orientation of membrane protrusions, lamellipodia	Tahinci and Symes, 2003
rac1	*Xenopus laevis*	Rho family small GTPase	Convergent extension	Intercalation, orientation of membrane protrusions, lamellipodia and filopodia	Tahinci and Symes, 2003
protein kinase Cδ	*Xenopus laevis*	Novel PKC, activated by diacylglycerol but not Ca^{2+}	Convergent extension	Mediolaterally-oriented cell elongation, intercalation	Kinoshita et al., 2003

processes that shape the zebrafish gastrula. It is our hope that continued efforts will not only contribute to our knowledge of these processes, but also provide important insight into how deregulation of pathways, such as canonical and noncanonical Wnt signaling, promotes tumor invasiveness and metastasis.

Acknowledgments

We thank members of our laboratory for many fruitful discussions and helpful comments on this chapter. J.R.J. acknowledges support by a National Institutes of Health Vascular Biology Training Grant (T32HL07751) and a fellowship from the American Cancer Society (PF-03-256-01-DDC). Work in the L.S.-K. lab is supported by NIH grants GM55101 and GM62283.

References

Adler, P.N. 2002. Planar signaling and morphogenesis in Drosophila. Dev. Cell. 2, 525–535.

Ataliotis, P., Symes, K., Chou, M.M., Ho, L., Mercola, M. 1995. PDGF signalling is required for gastrulation of Xenopus laevis. Development 121, 3099–3110.

Axelrod, J.D., Miller, J.R., Shulman, J.M., Moon, R.T., Perrimon, N. 1998. Differential recruitment of Disheveled provides signaling specificity in the planar cell polarity and Wingless signaling pathways. Genes Dev. 12, 2610–2622.

Bastock, R., Strutt, H., Strutt, D. 2003. Strabismus is asymmetrically localised and binds to Prickle and Disheveled during Drosophila planar polarity patterning. Development 130, 3007–3014.

Betchaku, T., Trinkaus, J.P. 1978. Contact relations, surface activity, and cortical microfilaments of marginal cells of the enveloping layer and of the yolk syncytial and yolk cytoplasmic layers of Fundulus before and during epiboly. J. Exp. Zool. 206, 381–426.

Bhanot, P., Brink, M., Samos, C.H., Hsieh, J.C., Wang, Y., Macke, J.P., Andrew, D., Nathans, J., Nusse, R. 1996. A new member of the frizzled family from Drosophila functions as a Wingless receptor. Nature 382, 225–230.

Bhattacharyya, R., Wedegaertner, P.B. 2000. Galpha 13 requires palmitoylation for plasma membrane localization, Rho-dependent signaling, and promotion of p115-RhoGEF membrane binding. J. Biol. Chem. 275, 14992–14999.

Boutros, M., Mlodzik, M. 1999. Disheveled: At the crossroads of divergent intracellular signaling pathways. Mech. Dev. 83, 27–37.

Cadigan, K.M., Nusse, R. 1997. Wnt signaling: A common theme in animal development. Genes Dev. 11, 3286–3305.

Carmany-Rampey, A., Schier, A.F. 2001. Single-cell internalization during zebrafish gastrulation. Curr. Biol. 11, 1261–1265.

Carreira-Barbosa, F., Concha, M.L., Takeuchi, M., Ueno, N., Wilson, S.W., Tada, M. 2003. Prickle 1 regulates cell movements during gastrulation and neuronal migration in zebrafish. Development 130, 4037–4046.

Chan, J., Mably, J.D., Serluca, F.C., Chen, J.N., Goldstein, N.B., Thomas, M.C., Cleary, J.A., Brennan, C., Fishman, M.C., Roberts, T.M. 2001. Morphogenesis of prechordal plate and notochord requires intact Eph/ephrin B signaling. Dev. Biol. 234, 470–482.

Choi, S.C., Han, J.K. 2002. Xenopus Cdc42 regulates convergent extension movements during gastrulation through Wnt/Ca2+ signaling pathway. Dev. Biol. 244, 342–357.

Chung, C.Y., Funamoto, S., Firtel, R.A. 2001. Signaling pathways controlling cell polarity and chemotaxis. Trends Biochem. Sci. 26, 557–566.

Concha, M.L., Adams, R.J. 1998. Oriented cell divisions and cellular morphogenesis in the zebrafish gastrula and neurula: A time-lapse analysis. Development 125, 983–994.

Cong, F., Schweizer, L., Chamorro, M., Varmus, H. 2003. Requirement for a nuclear function of beta-catenin in Wnt signaling. Mol. Cell. Biol. 23, 8462–8470.

Darken, R.S., Scola, A.M., Rakeman, A.S., Das, G., Mlodzik, M., Wilson, P.A. 2002. The planar polarity gene strabismus regulates convergent extension movements in Xenopus. EMBO J. 21, 976–985.

Devreotes, P., Janetopoulos, C. 2003. Eukaryotic chemotaxis: Distinctions between directional sensing and polarization. J. Biol. Chem. 278, 20445–20448.

Djiane, A., Riou, J., Umbhauer, M., Boucaut, J., Shi, D. 2000. Role of frizzled 7 in the regulation of convergent extension movements during gastrulation in Xenopus laevis. Development 127, 3091–3100.

Elul, T., Keller, R. 2000. Monopolar protrusive activity: A new morphogenic cell behavior in the neural plate dependent on vertical interactions with the mesoderm in Xenopus. Dev. Biol. 224, 3–19.

Erter, C.E., Wilm, T.P., Basler, N., Wright, C.V., Solnica-Krezel, L. 2001. Wnt8 is required in lateral mesendodermal precursors for neural posteriorization *in vivo*. Development 128, 3571–3583.

Fekany, K., Yamanaka, Y., Leung, T., Sirotkin, H.I., Topczewski, J., Gates, M.A., Hibi, M., Renucci, A., Stemple, D., Radbill, A., et al. 1999. The zebrafish bozozok locus encodes Dharma, a homeodomain protein essential for induction of gastrula organizer and dorsoanterior embryonic structures. Development 126, 1427–1438.

Feldman, B., Gates, M.A., Egan, E.S., Dougan, S.T., Rennebeck, G., Sirotkin, H.I., Schier, A.F., Talbot, W.S. 1998. Zebrafish organizer development and germ-layer formation require nodal-related signals. Nature 395, 181–185.

Firtel, R.A., Meili, R. 2000. Dictyostelium: A model for regulated cell movement during morphogenesis. Curr. Opin. Genet. Dev. 10, 421–427.

Fukuhara, S., Chikumi, H., Gutkind, J.S. 2000. Leukemia-associated Rho guanine nucleotide exchange factor (LARG) links heterotrimeric G proteins of the G(12) family to Rho. FEBS Lett. 485, 183–188.

Funamoto, S., Milan, K., Meili, R., Firtel, R.A. 2001. Role of phosphatidylinositol 3′ kinase and a downstream pleckstrin homology domain-containing protein in controlling chemotaxis in dictyostelium. J. Cell Biol. 153, 795–810.

Glickman, N.S., Kimmel, C.B., Jones, M.A., Adams, R.J. 2003. Shaping the zebrafish notochord. Development 130, 873–887.

Gonzalez, E., Fekany-Lee, K., Carmany-Rampey, A., Erter, C., Topczewski, J., Wright, C.V.E., Solnica-Krezel, L. 2000. Head and trunk development in zebrafish requires inhibition of Bmp signaling by bozozok and chordino. Genes Dev. 14, 3087–3092.

Goto, T., Keller, R. 2002. The planar cell polarity gene strabismus regulates convergence and extension and neural fold closure in Xenopus. Dev. Biol. 247, 165–181.

Gritsman, K., Zhang, J., Cheng, S., Heckscher, E., Talbot, W.S., Schier, A.F. 1999. The EGF-CFC protein one-eyed pinhead is essential for Nodal signaling. Cell 97, 121–132.

Gumbiner, B.M. 2000. Regulation of cadherin adhesive activity. J. Cell Biol. 148, 399–404.

Habas, R., Dawid, I.B., He, X. 2003. Coactivation of Rac and Rho by Wnt/Frizzled signaling is required for vertebrate gastrulation. Genes Dev. 17, 295–309.

Habas, R., Kato, Y., He, X. 2001. Wnt/Frizzled activation of Rho regulates vertebrate gastrulation and requires a novel Formin homology protein Daam1. Cell 107, 843–854.

Hammerschmidt, M., Mullins, M.C. 2002. Dorsoventral patterning in the zebrafish: Bone morphogenetic proteins and beyond. Results Probl. Cell Differ. 40, 72–95.

Hammerschmidt, M., Pelegri, F., Mullins, M.C., Kane, D.A., Brand, M., van Eeden, F.J., Furutani-Seiki, M., Granato, M., Haffter, P., Heisenberg, C.P., et al. 1996. Mutations affecting morphogenesis during gastrulation and tail formation in the zebrafish, Danio rerio. Development 123, 143–151.

Hart, M.J., Jiang, X., Kozasa, T., Roscoe, W., Singer, W.D., Gilman, A.G., Sternweis, P.C., Bollag, G. 1998. Direct stimulation of the guanine nucleotide exchange activity of p115 RhoGEF by Galpha13. Science 280, 2112–2114.

Haugh, J.M., Codazzi, F., Teruel, M., Meyer, T. 2000. Spatial sensing in fibroblasts mediated by 3′ phosphoinositides. J. Cell Biol. 151, 1269–1280.

Heisenberg, C.P., Brand, M., Jiang, Y.J., Warga, R.M., Beuchle, D., van Eeden, F.J., Furutani-Seiki, M., Granato, M., Haffter, P., Hammerschmidt, M., et al. 1996. Genes involved in forebrain development in the zebrafish, Danio rerio. Development 123, 191–203.

Heisenberg, C.P., Nusslein-Volhard, C. 1997. The function of silberblick in the positioning of the eye anlage in the zebrafish embryo. Dev. Biol. 184, 85–94.

Heisenberg, C.P., Tada, M. 2002. Zebrafish gastrulation movements: Bridging cell and developmental biology. Semin. Cell Dev. Biol. 13, 471–479.

Heisenberg, C.P., Tada, M., Rauch, G.J., Saude, L., Concha, M.L., Geisler, R., Stemple, D.L., Smith, J.C., Wilson, S.W. 2000. Silberblick/Wnt11 mediates convergent extension movements during zebrafish gastrulation. Nature 405, 76–81.

Hibi, M., Hirano, T., Dawid, I.B. 2002. Organizer formation and function. Results Probl. Cell Differ. 40, 48–71.

Ho, R.K., Kane, D.A. 1990. Cell-autonomous action of zebrafish *spt-1* mutation in specific mesodermal precursors. Nature 348, 728–730.

Jessen, J.R., Topczewski, J., Bingham, S., Sepich, D.S., Marlow, F., Chandrasekhar, A., Solnica-Krezel, L. 2002. Zebrafish trilobite identifies new roles for Strabismus in gastrulation and neuronal movements. Nat. Cell Biol. 4, 610–615.

Jimenez, C., Portela, R.A., Mellado, M., Rodriguez-Frade, J.M., Collard, J., Serrano, A., Martinez, A.C., Avila, J., Carrera, A.C. 2000. Role of the PI3K regulatory subunit in the control of actin organization and cell migration. J. Cell Biol. 151, 249–262.

Jin, T., Zhang, N., Long, Y., Parent, C.A., Devreotes, P.N. 2000. Localization of the G protein beta-gamma complex in living cells during chemotaxis. Science 287, 1034–1036.

Kane, D., Adams, R. 2002. Life at the edge: Epiboly and involution in the zebrafish. Results Probl. Cell Differ. 40, 117–135.

Kane, D.A., Hammerschmidt, M., Mullin, M.C., Maischein, H.-M., Brand, M., van Eeden, F.J.M., Furutani-Seiki, M., Granato, M., Haffter, P., Heisenberg, C.-P., et al. 1996. The zebrafish epiboly mutants. Development 123, 47–55.

Keller, R. 2002. Shaping the vertebrate body plan by polarized embryonic cell movements. Science 298, 1950–1954.

Keller, R., Davidson, L., Edlund, A., Elul, T., Ezin, M., Shook, D., Skoglund, P. 2000. Mechanisms of convergence and extension by cell intercalation. Philos. Trans. R Soc. Lond. B Biol. Sci. 355, 897–922.

Kilian, B., Mansukoski, H., Barbosa, F.C., Ulrich, F., Tada, M., Heisenberg, C.P. 2003. The role of Ppt/Wnt5 in regulating cell shape and movement during zebrafish gastrulation. Mech. Dev. 120, 467–476.

Kimmel, C.B., Ballard, W.W., Kimmel, S.R., Ullmann, B., Schilling, T.F. 1995. Stages of embryonic development of the zebrafish. Dev. Dyn. 203, 253–310.

Kimmel, C.B., Warga, R.M. 1987. Indeterminate cell lineage of the zebrafish embryo. Dev. Biol. 124, 269–280.

Kimmel, C.B., Warga, R.M., Kane, D.A. 1994. Cell cycles and clonal strings during formation of the zebrafish central nervous system. Development 120, 265–276.

Kinoshita, N., Iioka, H., Miyakoshi, A., Ueno, N. 2003. PKC delta is essential for Disheveled function in a noncanonical Wnt pathway that regulates Xenopus convergent extension movements. Genes Dev. 17, 1663–1676.

Kühl, M., Sheldahl, L.C., Malbon, C.C., Moon, R.T. 2000a. Ca(2+)/calmodulin-dependent protein kinase II is stimulated by Wnt and Frizzled homologs and promotes ventral cell fates in Xenopus. J. Biol. Chem. 275, 12701–12711.

Kühl, M., Sheldahl, L.C., Park, M., Miller, J.R., Moon, R.T. 2000b. The Wnt/Ca2+ pathway: A new vertebrate Wnt signaling pathway takes shape. Trends Genet. 16, 279–283.

Latinkic, B.V., Mercurio, S., Bennett, B., Hirst, E.M., Xu, Q., Lau, L.F., Mohun, T.J., Smith, J.C. 2003. Xenopus Cyr61 regulates gastrulation movements and modulates Wnt signalling. Development 130, 2429–2441.

Lele, Z., Bakkers, J., Hammerschmidt, M. 2001. Morpholino phenocopies of the swirl, snailhouse, somitabun, minifin, silberblick, and pipetail mutations. Genesis 30, 190–194.

Leung, T., Soll, I., Arnold, S.J., Kemler, R., Driever, W. 2003. Direct binding of Lef1 to sites in the boz promoter may mediate pre-midblastula-transition activation of boz expression. Dev. Dyn. 228, 424–432.

Liu, L., Chong, S.W., Balasubramaniyan, N.V., Korzh, V., Ge, R. 2002. Platelet-derived growth factor receptor alpha (pdgfr-alpha) gene in zebrafish embryonic development. Mech. Dev. 116, 227–230.

Makita, R., Mizuno, T., Kuroiwa, A., Koshida, S., Takeda, H. 1998. Zebrafish Wnt11: Pattern and regulation of the expression by the yolk cell and no tail activity. Mech. Dev. 71, 165–176.

Marlow, F., Topczewski, J., Sepich, D., Solnica-Krezel, L. 2002. Zebrafish rho kinase 2 acts downstream of Wnt11 to mediate cell polarity and effective convergence and extension movements. Curr. Biol. 12, 876–884.

Marlow, F., Zwartkruis, F., Malicki, J., Neuhauss, S.C., Abbas, L., Weaver, M., Driever, W., Solnica-Krezel, L. 1998. Functional interactions of genes mediating convergent extension, knypek and trilobite, during the partitioning of the eye primordium in zebrafish. Dev. Biol. 203, 382–399.

Marsden, M., DeSimone, D.W. 2003. Integrin-ECM interactions regulate cadherin-dependent cell adhesion and are required for convergent extension in Xenopus. Curr. Biol. 13, 1182–1191.

Medina, A., Reintsch, W., Steinbeisser, H. 2000. Xenopus frizzled 7 can act in canonical and non-canonical Wnt signaling pathways: Implications on early patterning and morphogenesis. Mech. Dev. 92, 227–237.

Meili, R., Ellsworth, C., Lee, S., Reddy, T.B., Ma, H., Firtel, R.A. 1999. Chemoattractant-mediated transient activation and membrane localization of Akt/PKB is required for efficient chemotaxis to cAMP in Dictyostelium. EMBO J. 18, 2092–2105.

Mlodzik, M. 2002. Planar cell polarization: Do the same mechanisms regulate Drosophila tissue polarity and vertebrate gastrulation? Trends Genet. 18, 564–571.

Montero, J.A., Kilian, B., Chan, J., Bayliss, P.E., Heisenberg, C.P. 2003. Phosphoinositide 3-kinase is required for process outgrowth and cell polarization of gastrulating mesendodermal cells. Curr. Biol. 13, 1279–1289.

Moon, R.T., Campbell, R.M., Christian, J.L., McGrew, L.L., Shih, J., Fraser, S. 1993. Xwnt-5A: A maternal Wnt that affects morphogenetic movements after overexpression in embryos of Xenopus laevis. Development 119, 97–111.

Myers, D., Sepich, D., Solnica-Krezel, L. 2002a. Convergence and extension in vertebrate gastrulae: Cell movements according to or in search of identity? Trends Genet. 18, 447.

Myers, D.C., Sepich, D.S., Solnica-Krezel, L. 2002b. Bmp activity gradient regulates convergent extension during zebrafish gastrulation. Dev. Biol. 243, 81–98.

Nusse, R., Varmus, H.E. 1982. Many tumors induced by the mouse mammary tumor virus contain a provirus integrated in the same region of the host genome. Cell 31, 99–109.

Oates, A.C., Lackmann, M., Power, M.A., Brennan, C., Down, L.M., Do, C., Evans, B., Holder, N., Boyd, A.W. 1999. An early developmental role for eph-ephrin interaction during vertebrate gastrulation. Mech. Dev. 83, 77–94.

Ohkawara, B., Yamamoto, T.S., Tada, M., Ueno, N. 2003. Role of glypican 4 in the regulation of convergent extension movements during gastrulation in Xenopus laevis. Development 130, 2129–2138.

Oishi, I., Suzuki, H., Onishi, N., Takada, R., Kani, S., Ohkawara, B., Koshida, I., Suzuki, K., Yamada, G., Schwabe, G.C., et al. 2003. The receptor tyrosine kinase Ror2 is involved in non-canonical Wnt5a/JNK signalling pathway. Genes Cells 8, 645–654.

Park, M., Moon, R.T. 2002. The planar cell-polarity gene stbm regulates cell behaviour and cell fate in vertebrate embryos. Nat. Cell Biol. 4, 20–25.

Penton, A., Wodarz, A., Nusse, R. 2002. A mutational analysis of disheveled in Drosophila defines novel domains in the disheveled protein as well as novel suppressing alleles of axin. Genetics 161, 747–762.

Penzo-Mendèz, A., Umbhauer, M., Djiane, A., Boucaut, J.C., Riou, J.F. 2003. Activation of Gbetagamma signaling downstream of Wnt-11/Xfz7 regulates Cdc42 activity during Xenopus gastrulation. Dev. Biol. 257, 302–314.

Rauch, G.J., Hammerschmidt, M., Blader, P., Schauerte, H.E., Strähle, U., Ingham, P.W., McMahon, A.P., Haffter, P. 1997. Wnt5 is required for tail formation in the zebrafish embryo. Cold Spring Harb. Symp. Quant. Biol. 62, 227–234.

Ridley, A.J. 2001. Rho proteins: Linking signaling with membrane trafficking. Traffic 2, 303–310.

Rothbächer, U., Laurent, M.N., Deardorff, M.A., Klein, P.S., Cho, K.W., Fraser, S.E. 2000. Disheveled phosphorylation, subcellular localization and multimerization regulate its role in early embryogenesis. EMBO J. 19, 1010–1022.

Ryu, S.L., Fujii, R., Yamanaka, Y., Shimizu, T., Yabe, T., Hirata, T., Hibi, M., Hirano, T. 2001. Regulation of dharma/bozozok by the Wnt pathway. Dev. Biol. 231, 397–409.

Sakai, T., Li, S., Docheva, D., Grashoff, C., Sakai, K., Kostka, G., Braun, A., Pfeifer, A., Yurchenco, P.D., Fassler, R. 2003. Integrin-linked kinase (ILK) is required for polarizing the epiblast, cell adhesion, and controlling actin accumulation. Genes. Dev. 17, 926–940.

Schmitz, A.A., Govek, E.E., Bottner, B., Van Aelst, L. 2000. Rho GTPases: Signaling, migration, and invasion. Exp. Cell Res. 261, 1–12.

Sepich, D.S., Myers, D.C., Short, R., Topczewski, J., Marlow, F., Solnica-Krezel, L. 2000. Role of the zebrafish trilobite locus in gastrulation movements of convergence and extension. Genesis 27, 159–173.

Servant, G., Weiner, O.D., Herzmark, P., Balla, T., Sedat, J.W., Bourne, H.R. 2000. Polarization of chemoattractant receptor signaling during neutrophil chemotaxis. Science 287, 1037–1040.

Sheldahl, L.C., Park, M., Malbon, C.C., Moon, R.T. 1999. Protein kinase C is differentially stimulated by Wnt and Frizzled homologs in a G-protein-dependent manner. Curr. Biol. 9, 695–698.

Sheldahl, L.C., Slusarski, D.C., Pandur, P., Miller, J.R., Kühl, M., Moon, R.T. 2003. Disheveled activates Ca2+ flux, PKC, and CamKII in vertebrate embryos. J. Cell Biol. 161, 769–777.

Shih, J., Keller, R. 1992a. Cell motility driving mediolateral intercalation in explants of *Xenopus laevis*. Development 116, 901–914.

Shih, J., Keller, R. 1992b. Patterns of cell motility in the organizer and dorsal mesoderm of *Xenopus laevis*. Development 116, 915–930.

Simon, A.R., Vikis, H.G., Stewart, S., Fanburg, B.L., Cochran, B.H., Guan, K.L. 2000. Regulation of STAT3 by direct binding to the Rac1 GTPase. Science 290, 144–147.

Slusarski, D.C., Corces, V.G., Moon, R.T. 1997a. Interaction of Wnt and a Frizzled homologue triggers G-protein-linked phosphatidylinositol signalling. Nature 390, 410–413.

Slusarski, D.C., Yang-Snyder, J., Busa, W.B., Moon, R.T. 1997b. Modulation of embryonic intracellular Ca2+ signaling by Wnt-5A. Dev. Biol. 182, 114–120.

Solnica-Krezel, L., Cooper, M.S. 2002. Cellular and genetic mechanisms of convergence and extension. Results Probl. Cell Differ. 40, 136–165.

Solnica-Krezel, L., Driever, W. 1994. Microtubule arrays of the zebrafish yolk cell: Organization and function during epiboly. Development 120, 2443–2455.

Solnica-Krezel, L., Stemple, D.L., Mountcastle-Shah, E., Rangini, Z., Neuhauss, S.C., Malicki, J., Schier, A.F., Stainier, D.Y., Zwartkruis, F., Abdelilah, S., Driever, W. 1996. Mutations affecting cell fates and cellular rearrangements during gastrulation in zebrafish. Development 123, 67–80.

Strähle, U., Jesuthasan, S. 1993. Ultraviolet irradiation impairs epiboly in zebrafish embryos: evidence for a microtubule-dependent mechanism of epiboly. Development 119, 909–919.

Strutt, D. 2003. Frizzled signalling and cell polarisation in Drosophila and vertebrates. Development 130, 4501–4513.

Sumanas, S., Ekker, S.C. 2001. Xenopus frizzled-7 morphant displays defects in dorsoventral patterning and convergent extension movements during gastrulation. Genesis 30, 119–122.

Sumanas, S., Kim, H.J., Hermanson, S., Ekker, S.C. 2001. Zebrafish frizzled-2 morphant displays defects in body axis elongation. Genesis 30, 114–118.

Sumanas, S., Strege, P., Heasman, J., Ekker, S.C. 2000. The putative Wnt receptor Xenopus frizzled-7 functions upstream of beta-catenin in vertebrate dorsoventral mesoderm patterning. Development 127, 1981–1990.

Symes, K., Mercola, M. 1996. Embryonic mesoderm cells spread in response to platelet-derived growth factor and signaling by phosphatidylinositol 3-kinase. Proc. Natl. Acad. Sci. USA 93, 9641–9644.

Tada, M., Smith, J.C. 2000. Xwnt11 is a target of Xenopus Brachyury: Regulation of gastrulation movements via Disheveled, but not through the canonical Wnt pathway. Development 127, 2227–2238.

Tahinci, E., Symes, K. 2003. Distinct functions of Rho and Rac are required for convergent extension during Xenopus gastrulation. Dev. Biol. 259, 318–335.

Takeuchi, M., Nakabayashi, J., Sakaguchi, T., Yamamoto, T.S., Takahashi, H., Takeda, H., Ueno, N. 2003. The prickle-related gene in vertebrates is essential for gastrulation cell movements. Curr. Biol. 13, 674–679.

Topczewski, J., Sepich, D.S., Myers, D.C., Walker, C., Amores, A., Lele, Z., Hammerschmidt, M., Postlethwait, J., Solnica-Krezel, L. 2001. The zebrafish glypican knypek controls cell polarity during gastrulation movements of convergent extension. Dev. Cell 1, 251–264.

Torres, M.A., Yang-Snyder, J.A., Purcell, S.M., DeMarais, A.A., McGrew, L.L., Moon, R.T. 1996. Activities of the Wnt-1 class of secreted signaling factors are antagonized by the Wnt-5A class and by a dominant negative cadherin in early Xenopus development. J. Cell Biol. 133, 1123–1137.

Tree, D.R., Shulman, J.M., Rousset, R., Scott, M.P., Gubb, D., Axelrod, J.D. 2002. Prickle mediates feedback amplification to generate asymmetric planar cell polarity signaling. Cell 109, 371–381.

Trinkaus, J.P. 1951. A study of the mechanism of epiboly in the egg of *Fundulus heteroclitus*. J. Exp. Zool. 118, 269–320.

Trinkaus, J.P. 1984. Mechanism of *Fundulus* epiboly–A current view. Amer. Zool. 24, 673–688.

Trinkaus, J.P. 1996. Ingression during early gastrulation of fundulus. Dev. Biol. 177, 356–370.

Trinkaus, J.P. 1998. Gradient in convergent cell movement during Fundulus gastrulation. J. Exp. Zool. 281, 328–335.

Trinkaus, J.P., Trinkaus, M., Fink, R.D. 1991. *In vivo* analysis of convergent cell movements in the germ ring of Fundulus. In: *Gastrulation, Movements, Patterns and Molecules* (R. Keller, W.H.J. Clark, F. Griffin, Eds.), New York and London: Plenum Press, pp. 121–134.

Trinkaus, J.P., Trinkaus, M., Fink, R.D. 1992. On the convergent cell movements of gastrulation in Fundulus. J. Exp. Zool. 261, 40–61.

Ulrich, F., Concha, M.L., Heid, P.J., Voss, E., Witzel, S., Roehl, H., Tada, M., Wilson, S.W., Adams, R.J., Soll, D.R., Heisenberg, C.P. 2003. Slb/Wnt11 controls hypoblast cell migration and morphogenesis at the onset of zebrafish gastrulation. Development 130, 5375–5384.

Veeman, M.T., Axelrod, J.D., Moon, R.T. 2003a. A second canon. Functions and mechanisms of beta-catenin-independent Wnt signaling. Dev. Cell. 5, 367–377.

Veeman, M.T., Slusarski, D.C., Kaykas, A., Louie, S.H., Moon, R.T. 2003b. Zebrafish prickle a modulator of noncanonical Wnt/Fz signaling, regulates gastrulation movements. Curr. Biol. 13, 680–685.

Vila-Coro, A.J., Rodriguez-Frade, J.M., Martin De Ana, A., Moreno-Ortiz, M.C., Martinez, A.C., Mellado, M. 1999. The chemokine SDF-1alpha triggers CXCR4 receptor dimerization and activates the JAK/STAT pathway. FASEB J. 13, 1699–1710.

Wallingford, J.B., Fraser, S.E., Harland, R.M. 2002. Convergent extension: The molecular control of polarized cell movement during embryonic development. Dev. Cell. 2, 695–706.

Wallingford, J.B., Rowning, B.A., Vogeli, K.M., Rothbacher, U., Fraser, S.E., Harland, R.M. 2000. Disheveled controls cell polarity during Xenopus gastrulation. Nature 405, 81–85.

Warga, R.M., Kimmel, C.B. 1990. Cell movements during epiboly and gastrulation in zebrafish. Development 108, 569–580.

Warga, R.M., Nusslein-Volhard, C. 1999. Origin and development of the zebrafish endoderm. Development 126, 827–838.

Watton, S.J., Downward, J. 1999. Akt/PKB localisation and 3′ phosphoinositide generation at sites of epithelial cell-matrix and cell-cell interaction. Curr. Biol. 9, 433–436.

Weiner, O.D. 2002. Regulation of cell polarity during eukaryotic chemotaxis: The chemotactic compass. Curr. Opin. Cell. Biol. 14, 196–202.

Wells, C.D., Liu, M.Y., Jackson, M., Gutowski, S., Sternweis, P.M., Rothstein, J.D., Kozasa, T., Sternweis, P.C. 2002. Mechanisms for reversible regulation between G13 and Rho exchange factors. J. Biol. Chem. 277, 1174–1181.

Westfall, T.A., Brimeyer, R., Twedt, J., Gladon, J., Olberding, A., Furutani-Seiki, M., Slusarski, D.C. 2003. Wnt-5/pipetail functions in vertebrate axis formation as a negative regulator of Wnt/beta-catenin activity. J. Cell. Biol. 162, 889–898.

Wharton, K.A., Jr. 2003. Runnin' with the Dvl: Proteins that associate with Dsh/Dvl and their significance to Wnt signal transduction. Dev. Biol. 253, 1–17.

Wilkie, T.M., Gilbert, D.J., Olsen, A.S., Chen, X.N., Amatruda, T.T., Korenberg, J.R., Trask, B.J., de Jong, P., Reed, R.R., Simon, M.I., et al. 1992. Evolution of the mammalian G protein alpha subunit multigene family. Nat. Genet. 1, 85–91.

Wilson, E.T., Cretekos, C.J., Helde, K.A. 1995. Cell mixing during early epiboly in the zebrafish embryo. Dev. Genet. 17, 6–15.

Winklbauer, R., Medina, A., Swain, R.K., Steinbeisser, H. 2001. Frizzled-7 signalling controls tissue separation during Xenopus gastrulation. Nature 413, 856–860.

Wodarz, A., Nusse, R. 1998. Mechanisms of Wnt signaling in development. Annu. Rev. Cell. Dev. Biol. 14, 59–88.

Xu, J., Wang, F., Van Keymeulen, A., Herzmark, P., Straight, A., Kelly, K., Takuwa, Y., Sugimoto, N., Mitchison, T., Bourne, H.R. 2003. Divergent signals and cytoskeletal assemblies regulate self-organizing polarity in neutrophils. Cell 114, 201–214.

Yagi, T., Takeichi, M. 2000. Cadherin superfamily genes: Functions, genomic organization, and neurologic diversity. Genes Dev. 14, 1169–1180.

Yamanaka, H., Moriguchi, T., Masuyama, N., Kusakabe, M., Hanafusa, H., Takada, R., Takada, S., Nishida, E. 2002. JNK functions in the non-canonical Wnt pathway to regulate convergent extension movements in vertebrates. EMBO Rep. 3, 69–75.

Yamashita, S., Miyagi, C., Carmany-Rampey, A., Shimizu, T., Fujii, R., Schier, A.F., Hirano, T. 2002. Stat3 Controls Cell Movements during Zebrafish Gastrulation. Dev. Cell. 2, 363–375.

Yap, A.S., Kovacs, E.M. 2003. Direct cadherin-activated cell signaling: A view from the plasma membrane. J. Cell. Biol. 160, 11–16.

Yeo, S.Y., Little, M.H., Yamada, T., Miyashita, T., Halloran, M.C., Kuwada, J.Y., Huh, T.L., Okamoto, H. 2001. Overexpression of a slit homologue impairs convergent extension of the mesoderm and causes cyclopia in embryonic zebrafish. Dev. Biol. 230, 1–17.

Index

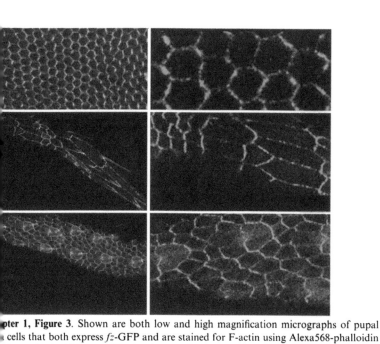

pter 1, Figure 3. Shown are both low and high magnification micrographs of pupal
cells that both express *fz*-GFP and are stained for F-actin using Alexa568-phalloidin
shown in the high magnification images. In these micrographs, *fz*-GFP is visualized
ibody staining. Note the distinctive zigzag *fz* pattern in the wing and that a similar
sta and leg, albeit modified to fit the cellular geometry of these cell types. In all of these
P is found in an uneven and clumpy pattern in the regions where it accumulates. Note
s the actin-rich hair is formed at the edge of the cell, perhaps in direct contact with the
the arista however, the laterals form some distance from the accumulated *fz*-GFP.
ccumulated asymmetrically in the developing bristle socket cells on the leg. The arista
modified versions of images from He and Adler, 2002 (4).

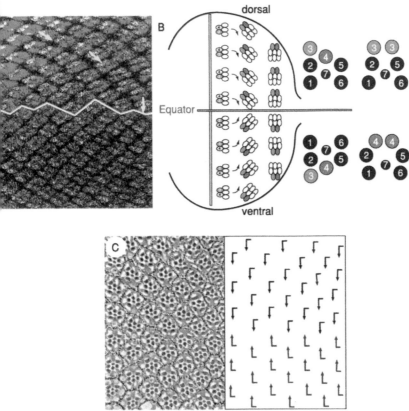

dzik, Chapter 2, Figure 1. Planar cell polarity features in the *Drosophila* eye. (A) Partial view of a
eye imaginal disc demonstrating the regularity of polarity establishment. Anterior is left and
this and all subsequent figures. Ommatidial clusters are marked with anti-Elav (green; labeling all
tors) and *svp-lacZ* (magenta; *svp* is expressed initially in R3/R4—see left side of panel—and later
R6 at weaker levels). The morphogenetic furrow is on the left side adjacent to field shown. The
and degree of rotation of some dorsal ommatidial preclusters is highlighted with yellow arrows;
marks the equator. (B) Schematic drawing of third instar larval eye imaginal disc, with the
etic furrow (M; yellow) and the D/V midline (the equator; gray) indicated. Initially, ommatidial
are symmetrical and organized in the A/P axis. Subsequently they rotate 90° with respect to the
the end of this process chirality is established by the positions of R3 and R4. Right side: schematic
n of chiral organization of dorsal and ventral adult ommatidia; in addition to the chiral forms,
l clusters with R3/R3 or R4/R4 cell pairs (as often found in PCP mutant tissue) are also shown. R3
hlighted in green and R4s in magenta. (C) Tangential section of a wild-type adult eye (left panel)
pective schematic presentation (right panel). The dorsal and ventral ommatidial arrangements are
by black and red arrows, respectively. Note the very regular ommatidial arrangement and the
or image symmetry between the dorsal and ventral halves (17).

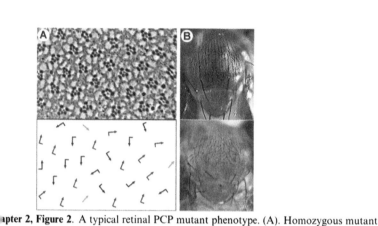

apter 2, **Figure 2**. A typical retinal PCP mutant phenotype. (A). Homozygous mutant s schematic presentation in lower panel. Arrows are drawn as in Fig. 1 with black and ting dorsal and ventral ommatidial chirality, respectively, and symmetrical R3/R3 or re represented by green arrows. The equatorial arrangement is lost with a random chiralities (black and red arrows), and the presence of several symmetrical clusters of types (green arrows), compare to Fig. 1C for regular wild-type arrangement. (B) PCP sensory touch bristles on the dorsal notum for comparison (upper panel: *wild type*; Note that in both tissues the orientation of the respective neural sensory units is he body axes (20).

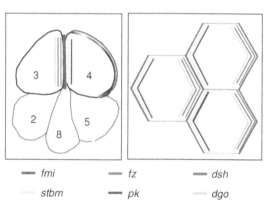

fmi   fz   dsh

stbm   pk   dgo

apter 2, **Figure 4**. Subcellular localization of PCP factors in the eye and wing. In the subcellular localization differences of PCP proteins are restricted to the R3/R4 anel), whereas in the wing they are apparent in all cells of the differentiating wing field parison). Initially, all PCP proteins are localized around the whole apical cortex (not g their interactions become restricted to either the R3 side or the R4 side of the R3/R4 color code for the PCP factors is as shown. Note that *fmi* (purple) localizes to both boundary within the DV axis (and both sides of each wing cell boundary in the

The functional significance for the enrichment of all PCP factors around the R4 hown in rainbow colors) is not known, and there is no genetic evidence for a *sh*, or *dgo* in R4. The localization of *dsh* and *pk* is deduced from their genetic l localizations in wing cells and stable physical interactions with with a fe (f. 44)

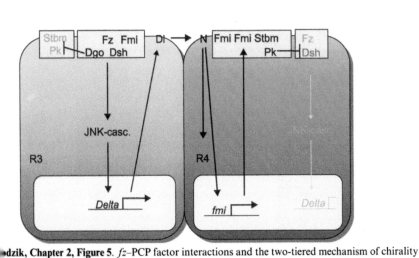

dzik, Chapter 2, Figure 5. *fz*–PCP factor interactions and the two-tiered mechanism of chirality within the R3/R4 pair. An initial difference in *fz–dsh*–PCP signaling levels, as generated ~e interactions between *fz–dsh* and the other PCP factors, is amplified by the transcriptional ~n of *Delta* and the subsequent activation of Notch signaling in R4. Notch activation in turn ~lta transcription in R4, further increasing the *fz*–PCP signaling difference between the R3 and ~d thus creating a very solid binary cell fate decision. In mutants of PCP factors the initial ~ of *Delta* upregulation is lost and a stochastic Delta/Notch interaction generates randomly the ~ fates and chirality. Within the PCP gene cassette the role of *stbm/Vang* and *prickle* (*pk*) is ~–*dsh* signaling. Thus *stbm* and *pk* are required mainly in the R4 cell where they antagonize the ~thway. *fmi* appears to be required to stabilize both membrane-associated complexes, although ~isms of its action remains unknown as no physical interactions between *fmi* and any other PCP ~ been reported. The role of Diego is also not yet clear, except that it is positively required for ~naling. See text for details (30).

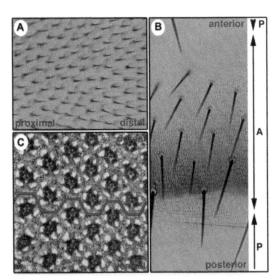

~tt and David Strutt, Chapter 3, Figure 2. Planar polarity in adult *Drosophila* tissues. (A) Each

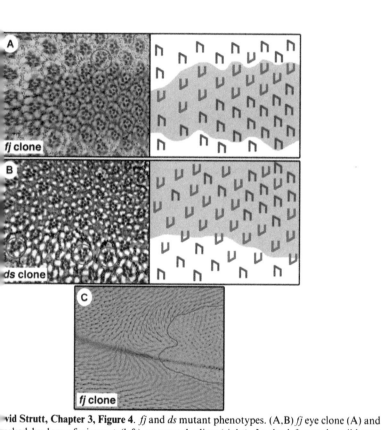

vid Strutt, Chapter 3, Figure 4. *fj* and *ds* mutant phenotypes. (A,B) *fj* eye clone (A) and
arked by loss of pigment (left) or grey shading (right). In the left panels, wild-type
ersed DV polarity are circled, whilst in the right panels all ommatidia with reversed
in red. Note that several rows of ommatidia on the polar side of the *fj* clone have
n (A); whilst inside the *ds* clone ommatidia have randomized polarity and
the equatorial boundary (B). (C) *fj* wing clone, mutant tissue marked by the hair
shavenoid, and outlined in red. Wild-type hairs proximal to the clone point away from

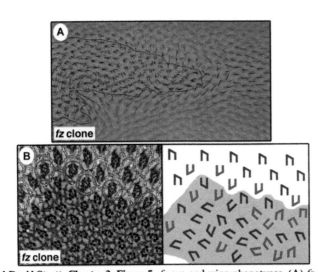

t and David Strutt, Chapter 3, Figure 5. *fz* eye and wing phenotypes. (A) *fz* clone in the wing, ...e marked by the *multiple wing hairs* gene, and outlined in red. Hairs distally and laterally to ...int towards the clone. (B) *fz* eye clone, marked by loss of pigment (left) or grey shading (right). ...one ommatidia have randomized polarity and orientation, and some wild-type ommatidia on ...one boundary have reversed DV polarity (circled in red). Dorsal-type ommatidia are in blue ...-type in red (50).

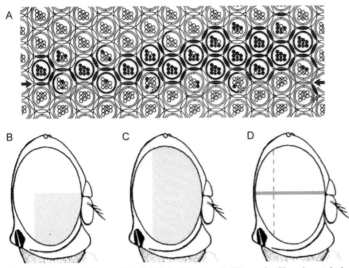

, Janghoo Lim and Kwang-Wook Choi, Chapter 4, Figure 2. Clonal restrictions in the eye. ...of mosaic clones in the adult eye. (A) Homozygous w^- clone (shown in black) in the equator ...saic clones were generated by X-ray-induced mitotic recombination during first instar larval ...atidia at the clone border consist of both w^+ and w^- photoreceptors, indicating that the eight ...tors in each ommatidium are not clonally related. Note that most w^- ommatidia are located in ...ide but a few w^- cells marked by red asterisks can be found in the ventral ommatidia below the ...rows). Adapted from Ready et al. (1976). (B–D) Representative large homozygous M^+/M^+ ...pink area). (B) This clone shows straight clone boundary at the dorsal ventral midline. (C) This

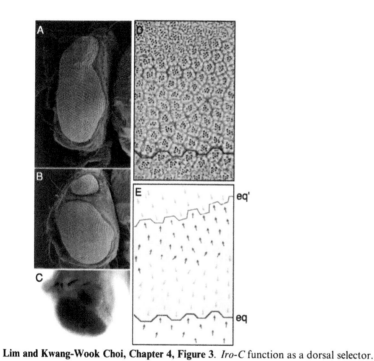

Lim and Kwang-Wook Choi, Chapter 4, Figure 3. *Iro-C* function as a dorsal selector. noving *Iro-C* activity in eye development. In all panels, anterior is to the left and Adult eyes harboring dorsal *Iro-C* clones. Mutant tissue is genetically labeled by the :aring as a pigmentless tissue against the red-pigmented wild-type tissue. (D) Section rying a dorsal (upper *white* tissue) and a ventral (lower *white* tissue) clone. (E) ation of the ommatidial polarity of the eye in (D). Dorsal (blue) and ventral (red) is represented by arrows. The equator (eq) is outlined by a thick blue line in D and E. orsal *Iro-C* clone defines an ectopic equator (eq', in E). The ventrally located clone otypic effect. Adapted from Cavodeassi et al. (2000) (68).

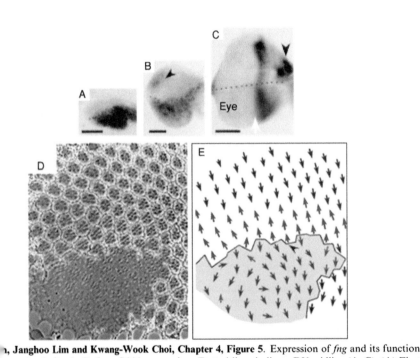

n, Janghoo Lim and Kwang-Wook Choi, Chapter 4, Figure 5. Expression of *fng* and its function on of polarity A–C, *fng* mRNA expression. Dotted lines indicate DV midline (A–C). (A) First (B) late second instar disc, (C) Late third instar disc. Arrowheads indicate weak *fng* expression domain (B) and expression in the ocelli region (C), respectively. Equatorial region and metic furrow are marked with black and white arrows, respectively (C). Scale bars: a, 5 μm; b, 45 μm. (D) Section of *fng* mosaic eye showing polarity reversals by ventral *fng⁻* clones (D). (E) presentaion of (D). *fng⁻* clones are colored green. Dorsal and ventral trapezoids are indicated blue arrows, respectively. *fng⁻* clone near the ventral margin shows polarity reversals in 3–4 matidia outside the clone border. An ommatidium with abnormal rotation but normal chirality with a black arrowhead. Anterior is to the right. Adapted from Cho and Choi (1998) (71).

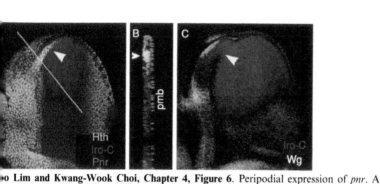

o Lim and Kwang-Wook Choi, Chapter 4, Figure 6. Peripodial expression of *pnr*. A
[a]nnal imaginal disc shows expression of GFP reporter of *pnr* expression driven by *pnr*-
[a]lso shows the expression of *hth* and *rF209*, a β-galactosidase reporter for *Iro-C*. *pnr* is
[ ]d in the dorsal peripodial cells. *Iro-C-lacZ* is expressed in the dorsal half of the eye
[ ]cted in the peripodial membrane. (B) Cross section of the disc in (A) at the angle
[ ] shows that *pnr* expression is restricted in the peripodial membrane. *Iro-C-lacZ* shows
[ ]*nr*. *hth* is also expressed in the peripodial membrane but does not overlap with *pnr*.
[a]ud and Casares (2000) (74).

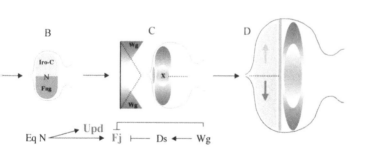

	Late L1-mid L2	Early L3	Mid-late L3
[ ]f	Establishment of DV pattern, Notch activation	Wg and secondary signal gradients, Furrow initiation	Differentiation, transduction of polarity signal

o Lim and Kwang-Wook Choi, Chapter 4, Figure 8. Schematic presentation of DV
[a] eye disc. The top panel shows a simplified model of gene functions involved in
[ ]ing larval eye development. (A) Mid-late first instar stage. Pnr expressed from
[a]l cells induces Wg expression. Secreted Wg allows expression of Iro-C genes in the
[ ]n. L and Ser are required for growth of the ventral domain. (B) Late first- to mid-
[ ]Iro-C and Fng domains are established and N is activated at the DV border. (C) Early
[ ]Vg secreted from the DV polar regions generate pole-to-equator gradient. Opposite
[ ]dients of secreted signaling molecules such as Upd and Fj may be formed from the
[ ] N activity at the DV border. The equator-to-pole gradient of Fj may be reinforced
[ ]m component. Upd and Fj are involved in generation of secondary polarity signal X
[ ]ting photoreceptor clusters respond by activating a signal transduction pathway to
[ ] polarity (step D) (82).

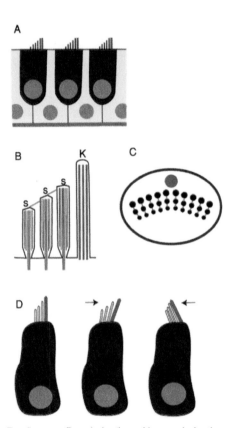

Resting   Depolarization   Hyperpolarization

Doub, Mireille Montcouquiol and Matthew W. Kelley, Chapter 6, Figure 1. Anatomy of
sory hair cell epithelia. (A) Cross-section of a typical hair cell sensory epithelium. The
is pseudostratified, with hair cells (blue) located near the lumenal surface and supporting cells
sitioned on the basement membrane (gray). Supporting cells extend lumenal projections that
 between each hair cell. A stereociliary bundle (black and red) extends from the lumenal
each hair cell. (B) Cross-section through the center of a stereociliary bundle. The bundle
multiple rows of stereocilia (S) and a single kinocilium (K). The core of each stereocilium
multiple actin filaments (gray) arranged in parallel. Note that stereocilia narrow at the base to
 point. Stereocilia are arranged in a staircase pattern such that the talles stereocilia are located
he kinocilium. The top of each stereocilium is connected to the side of the adjacent taller
 by a filamentous strand referred to as a tip link (green). There are other links located on the
h stereocilium but these have not been illustrated for clarity. The core of the single kinocilium
d of a 9 + 2 arrangement of microtubules (red) and there are no tip links between the stereocilia
ium. (C) View of a stereociliary bundle at the lumenal surface of a hair cell. Stereocilia are
 rows to create a "V" or rounded shape (not shown) with the kinocilium located at the vertex
dle. Tip links (green) extend between stereocilia in each row. (D) Illustration of hair cell
n. With the stereociliary bundle at rest, hair cells maintain a resting negative potential.
of the bundle in the direction of the kinocilium (red) leads to a rapid depolarization of the cell.
, deflection in the opposite direction results in cellular hyperpolarization. Deflections

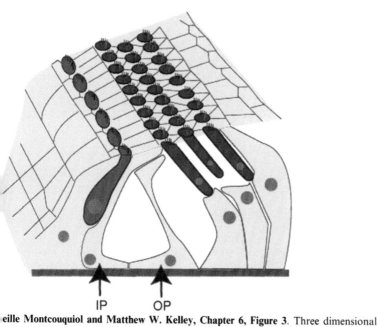

eille Montcouquiol and Matthew W. Kelley, Chapter 6, Figure 3. Three dimensional ory epithelium of the mammalian cochlea (also referred to as the organ of Corti). Hair to a single row of inner hair cells (green) and three rows of outer hair cells (blue). low) are specialized as well, with the most striking change being the formation of the a fluid-filled structure that forms between the inner and outer pillar cells (IP and OP).

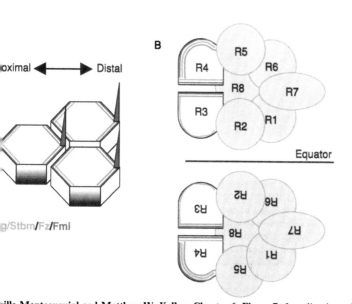

eille Montcouquiol and Matthew W. Kelley, Chapter 6, Figure 7. Localization of core eneration of polarity. (A) In the fly wing a homogenous population of cells each calized hair through differential localization of *vang, stbm, fz,* and *fmi*. While the

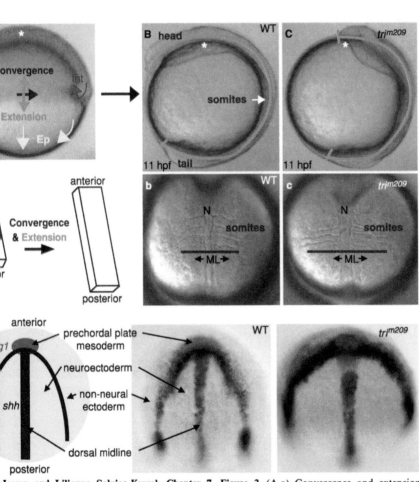

Jessen and Lilianna Solnica-Krezel, Chapter 7, Figure 3. (A,a) Convergence and extension ...s contribute significantly to shaping the early embryo by narrowing tissues mediolaterally while ...usly elongating them from head to tail. (B,b and C,c) In *tri* mutant embryos, impairment of ...ce and extension cell movements result in an embryo that has a shortened anteroposterior axis ...s) and is broadened mediolaterally (ML). (D) A typical whole-mount *in situ* hybridization ...emonstrating how the use of combinations of gene expression probes that mark specific ... tissues can suggest disrupted convergence and extension movements. Note that the homozygous ... embryo exhibits posteriorly shifted prechordal plate mesoderm (here, labeled in red with *hgg1*) ... tissue (here, labeled with *shh*), and a mediolaterally broadened neural plate (here, the border ...eural and non-neural ectoderm is labeled with *dlx3*). While *in situ* analyses are informative, they ...eal anything about the underlying cellular defects. The asterisk denotes the animal pole (137).

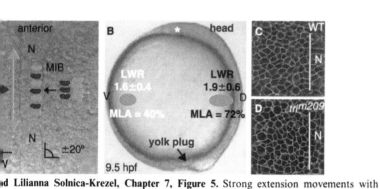

d Lilianna Solnica-Krezel, Chapter 7, Figure 5. Strong extension movements with
characterize the dorsal domain of convergence and extension. (A) Mediolateral
ors (MIB) underlying anteroposterior extension of axial tissues may be limited to the
zebrafish embryos. (A and B) Cell elongation and mediolateral-orientation are
aviors to examine in the dorsal domain. Asterisk denotes animal pole. (C and D) In *tri*
raxial ectodermal cells are rounder and less biased in their orientation than wild-type
eled with membrane-targeted GFP and live embryos were analyzed using confocal

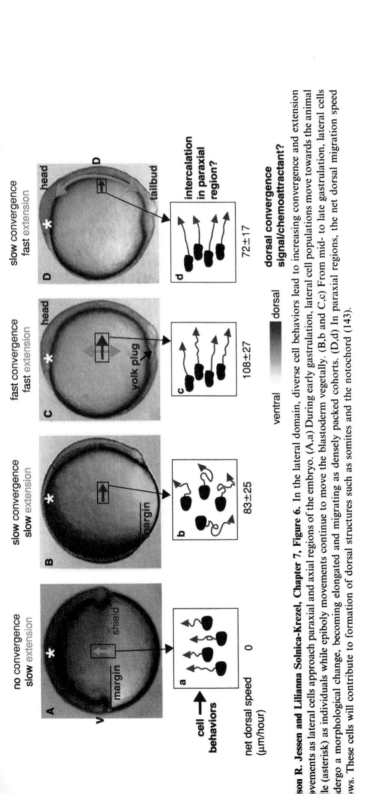

son R. Jessen and Lilianna Solnica-Krezel, Chapter 7, Figure 6. In the lateral domain, diverse cell behaviors lead to increasing convergence and extension vements as lateral cells approach paraxial and axial regions of the embryo. (A,a) During early gastrulation, lateral cell populations move towards the animal le (asterisk) as individuals while epiboly movements continue to move the blastoderm vegetally. (B,b and C,c) From mid- to late gastrulation, lateral cells dergo a morphological change, becoming elongated and migrating as densely packed cohorts. (D,d) In paraxial regions, the net dorsal migration speed ws. These cells will contribute to formation of dorsal structures such as somites and the notochord (143).

Printed and bound by CPI Group (UK) Ltd, Croydon, CR0 4YY

08/05/2025

01864966-0008